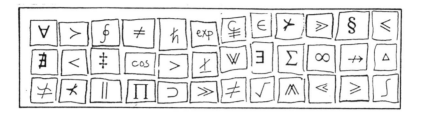

MATHEMATICAL WRITING

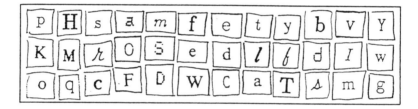

by Donald E. Knuth, Tracy Larrabee, and Paul M. Roberts

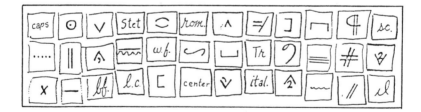

MAA Notes Series

The MAA Notes Series, started in 1982, addresses a broad range of topics and themes of interest to all who are involved with undergraduate mathematics. The volumes in this series are readable, informative, and useful, and help the mathematical community keep up with developments of importance to mathematics.

Editorial Board
Warren Page, Chair

Donald W. Bushaw   Richard K. Guy
Paul J. Campbell   Frederick Hoffman
Jane M. Day   David A. Smith

1. Problem Solving in the Mathematics Curriculum,
   Committee on the Undergraduate Teaching of Mathematics, Alan Schoenfeld, Editor.

2. Recommendations on the Mathematical Preparation of Teachers,
   CUPM Panel on Teacher Training.

3. Undergraduate Mathematics Education in the People's Republic of China,
   Lynn A. Steen, Editor.

4. Notes on Primality Testing and Factoring,
   by *Carl Pomerance.*

5. American Perspectives on the Fifth International Congress on Mathematical Education,
   Warren Page, Editor.

6. Toward a Lean and Lively Calculus,
   Ronald Douglas, Editor.

7. Undergraduate Programs in the Mathematical and Computer Sciences: 1985–86,
   D. J. Albers, R. D. Anderson, D. O. Loftsgaarden, Editors.

8. Calculus for a New Century,
   Lynn A. Steen, Editor.

9. Computers and Mathematics: The Use of Computers in Undergraduate Instruction,
   D. A. Smith, G. J. Porter, L. C. Leinbach, and R. H. Wenger, Editors.

10. Guidelines for the Continuing Mathematical Education of Teachers,
    Committee on the Mathematical Education of Teachers.

11. Keys to Improved Instruction by Teaching Assistants and Part-Time Instructors,
    Committee on Teaching Assistants and Part-Time Instructors, Bettye Anne Case, Editor.

12. The Use of Calculators in the Standardized Testing of Mathematics,
    John Kenelly, Editor.

13. Reshaping College Mathematics,
    Lynn A. Steen, Editor.

14. Mathematical Writing,
    by *Donald E. Knuth, Tracy Larrabee, and Paul M. Roberts.*

15. Discrete Mathematics in the First Two Years,
    Anthony Ralston, Editor.

16. Using Writing to Teach Mathematics,
    Andrew Sterrett, Editor.

17. Priming the Calculus Pump: Innovations and Resources,
    Thomas W. Tucker, Editor.

18. Models for Undergraduate Research in Mathematics,
    Lester Senechal, Editor.

Second Printing
© 1989 by the Mathematical Association of America
Library of Congress number: 89-062390
ISBN 0-88385-063-X
Printed in the United States of America
A version of these notes was previously available on a limited basis as

Mathematical Writing

by

Donald E. Knuth, Tracy Larrabee, and Paul M. Roberts

This report is based on a course of the same name given at Stanford University during autumn quarter, 1987. Here's the catalog description:

CS 209. Mathematical Writing—Issues of technical writing and the effective presentation of mathematics and computer science. Preparation of theses, papers, books, and "literate" computer programs. A term paper on a topic of your choice; this paper may be used for credit in another course.

The first three lectures were a "minicourse" that summarized the basics. About two hundred people attended those three sessions, which were devoted primarily to a discussion of the points in §1 of this report. An exercise (§2) and a suggested solution (§3) were also part of the minicourse.

The remaining 28 lectures covered these and other issues in depth. We saw many examples of "before" and "after" from manuscripts in progress. We learned how to avoid excessive subscripts and superscripts. We discussed the documentation of algorithms, computer programs, and user manuals. We considered the process of refereeing and editing. We studied how to make effective diagrams and tables, and how to find appropriate quotations to spice up a text. Some of the material duplicated some of what would be discussed in writing classes offered by the English department, but the vast majority of the lectures were devoted to issues that are specific to mathematics and/or computer science.

Guest lectures by Herb Wilf (University of Pennsylvania), Jeff Ullman (Stanford), Leslie Lamport (Digital Equipment Corporation), Nils Nilsson (Stanford), Mary-Claire van Leunen (Digital Equipment Corporation), Rosalie Stemer (San Francisco Chronicle), and Paul Halmos (University of Santa Clara), were a special highlight as each of these outstanding authors presented their own perspectives on the problems of mathematical communication.

This report contains transcripts of the lectures and copies of various handouts that were distributed during the quarter. We think the course was able to clarify a surprisingly large number of issues that play an important part in the life of every professional who works in mathematical fields. Therefore we hope that people who were unable to attend the course might still benefit from it, by reading this summary of what transpired.

The authors wish to thank Phyllis Winkler for the first-rate technical typing that made these notes possible.

Caveat: These are transcripts of lectures, not a polished set of essays on the subject. Some of the later lectures refer to mistakes in the notes of earlier lectures; we have decided to correct some (but not all) of those mistakes before printing this report. References to such no-longer-existent blunders might be hard to understand. Understand?

Videotapes of the class sessions are kept in the Mathematical & Computer Sciences Library at Stanford.

The preparation of this report was supported in part by NSF grant CCR-8610181.

A Grook by Piet Hein, cut into slate at David Kindersley's Workshop, Cambridge, England in 1988 for Donald E. Knuth. See page 49 and the material on quotations in this volume. The boundary of the figure is a superellipse.

Table of Contents

§1. Minicourse on technical writing 1
§2. An exercise on technical writing 7
§3. An answer to the exercise 8
§4. Comments on student answers (1) 9
§5. Comments on student answers (2) 11
§6. Preparing books for publication (1) 14
§7. Preparing books for publication (2) 15
§8. Preparing books for publication (3) 18
§9. Handy reference books 19
§10. Presenting algorithms 20
§11. Literate Programming (1) 22
§12. Literate Programming (2) 26
§13. User manuals 28
§14. Galley proofs 30
§15. Refereeing (1) 31
§16. Refereeing (2) 34
§17. Hints for Referees 36
§18. Illustrations (1) 37
§19. Illustrations (2) 40
§20. Homework: Subscripts and superscripts 40
§21. Homework: Solutions 43
§22. Quotations 47
§23. Scientific American Saga (1) 49
§24. Scientific American Saga (2) 51
§25. Examples of good style 54
§26. Mary-Claire van Leunen on 'hopefully' 57
§27. Herb Wilf on Mathematical Writing 59
§28. Wilf's first extreme 61
§29. Wilf's other extreme 62
§30. Jeff Ullman on Getting Rich 66
§31. Leslie Lamport on Writing Papers 69
§32. Lamport's handout on unnecessary prose 71
§33. Lamport's handout on styles of proof 72
§34. Nils Nilsson on Art and Writing 73
§35. Mary-Claire van Leunen on Calisthenics (1) 77
§36. Mary-Claire's handout on Composition Exercises 81
§37. Comments on student work 89
§38. Mary-Claire van Leunen on Which vs. That 93
§39. Mary-Claire van Leunen on Calisthenics (2) 98
§40. Computer aids to writing 100
§41. Rosalie Stemer on Copy Editing 102
§42. Paul Halmos on Mathematical Writing 106
§43. Final truths 112

§1. Notes on Technical Writing

Stanford's library card catalog refers to more than 100 books about technical writing, including such titles as *The Art of Technical Writing*, *The Craft of Technical Writing*, *The Teaching of Technical Writing*. There is even a journal devoted to the subject, the *IEEE Transactions on Professional Communication*, published since 1958. The American Chemical Society, the American Institute of Physics, the American Mathematical Society, and the Mathematical Association of America have each published "manuals of style." The last of these, *Writing Mathematics Well* by Leonard Gillman, is one of the required texts for CS 209.

The nicest little reference for a quick tutorial is *The Elements of Style*, by Strunk and White (Macmillan, 1979). Everybody should read this 85-page book, which tells about English prose writing in general. But it isn't a required text—it's merely recommended.

The other required text for CS 209 is *A Handbook for Scholars* by Mary-Claire van Leunen (Knopf, 1978). This well-written book is a real pleasure to read, in spite of its unexciting title. It tells about footnotes, references, quotations, and such things, done correctly instead of the old-fashioned "op. cit." way.

Mathematical writing has certain peculiar problems that have rarely been discussed in the literature. Gillman's book refers to the three previous classics in the field: An article by Harley Flanders, *Amer. Math. Monthly*, 1971, pp. 1–10; another by R. P. Boas in the same journal, 1981, pp. 727–731. There's also a nice booklet called *How to Write Mathematics*, published by the American Mathematical Society in 1973, especially the delightful essay by Paul R. Halmos on pp. 19–48.

The following points are especially important, in your instructor's view:

1. Symbols in different formulas must be separated by words.

    Bad: Consider S_q, $q < p$.

    Good: Consider S_q, where $q < p$.

2. Don't start a sentence with a symbol.

    Bad: $x^n - a$ has n distinct zeroes.

    Good: The polynomial $x^n - a$ has n distinct zeroes.

3. Don't use the symbols $\therefore$, $\Rightarrow$, $\forall$, $\exists$, $\ni$; replace them by the corresponding words. (Except in works on logic, of course.)

4. The statement just preceding a theorem, algorithm, etc., should be a complete sentence or should end with a colon.

    Bad: We now have the following
    Theorem. $H(x)$ is continuous.

This is bad on three counts, including rule 2. It should be rewritten, for example, like this:

   Good: We can now prove the following result.
   Theorem. The function $H(x)$ defined in (5) is continuous.

Even better would be to replace the first sentence by a more suggestive motivation, tying the theorem up with the previous discussion.

5. The statement of a theorem should usually be self-contained, not depending on the assumptions in the preceding text. (See the restatement of the theorem in point 4.)

6. The word "we" is often useful to avoid passive voice; the "good" first sentence of example 4 is much better than "The following result can now be proved." But this use of "we" should be used in contexts where it means "you and me together", *not* a formal equivalent of "I". Think of a dialog between author and reader.

    In most technical writing, "I" should be avoided, unless the author's persona is relevant.

7. There is a definite rhythm in sentences. Read what you have written, and change the wording if it does not flow smoothly. For example, in the text *Sorting and Searching* it was sometimes better to say "merge patterns" and sometimes better to say "merging patterns". There are many ways to say "therefore", but often only one has the correct rhythm.

8. Don't omit "that" when it helps the reader to parse the sentence.

     Bad: Assume A is a group.
     Good: Assume that A is a group.

    The words "assume" and "suppose" should usually be followed by "that" unless another "that" appears nearby. But *never* say "We have that $x = y$," say "We have $x = y$." And avoid unnecessary padding "because of the fact that" unless you feel that the reader needs a moment to recuperate from a concentrated sequence of ideas.

9. Vary the sentence structure and the choice of words, to avoid monotony. But use parallelism when parallel concepts are being discussed. For example (Strunk and White #15), don't say this:

    > Formerly, science was taught by the textbook method, while now the laboratory method is employed.

    Rather:

    > Formerly, science was taught by the textbook method; now it is taught by the laboratory method.

    Avoid words like "this" or "also" in consecutive sentences; such words, as well as unusual or polysyllabic utterances, tend to stick in a reader's mind longer than other words, and good style will keep "sticky" words spaced well apart. (For example, I'd better not say "utterances" any more in the rest of these notes.)

10. Don't use the style of homework papers, in which a sequence of formulas is merely listed. Tie the concepts together with a running commentary.

11. Try to state things twice, in complementary ways, especially when giving a definition. This reinforces the reader's understanding. (Examples, see §2 below: N^n is defined twice, A_n is described as "nonincreasing", $L(C, P)$ is characterized as the smallest subset of a certain type.) All variables must be defined, at least informally, when they are first introduced.

12. Motivate the reader for what follows. In the example of §2, Lemma 1 is motivated by the fact that its converse is true. Definition 1 is motivated only by decree; this is somewhat riskier.

    Perhaps the most important principle of good writing is to keep the reader uppermost in mind: What does the reader know so far? What does the reader expect next and why?

    When describing the work of other people it is sometimes safe to provide motivation by simply stating that it is "interesting" or "remarkable"; but it is best to let the results speak for themselves or to give *reasons* why the things seem interesting or remarkable.

    When describing your own work, be humble and don't use superlatives of praise, either explicitly or implicitly, even if you are enthusiastic.

13. Many readers will skim over formulas on their first reading of your exposition. Therefore, your sentences should flow smoothly when all but the simplest formulas are replaced by "blah" or some other grunting noise.

14. Don't use the same notation for two different things. Conversely, use consistent notation for the same thing when it appears in several places. For example, don't say "A_j for $1 \leq j \leq n$" in one place and "A_k for $1 \leq k \leq n$" in another place unless there is a good reason. It is often useful to choose names for indices so that i varies from 1 to m and j from 1 to n, say, and to stick to consistent usage. Typographic conventions (like lowercase letters for elements of sets and uppercase for sets) are also useful.

15. Don't get carried away by subscripts, especially when dealing with a set that doesn't need to be indexed; set element notation can be used to avoid subscripted subscripts. For example, it is often troublesome to start out with a definition like "Let $X = \{x_1, \ldots, x_n\}$" if you're going to need subsets of X, since the subset will have to defined as $\{x_{i_1}, \ldots, x_{i_m}\}$, say. Also you'll need to be speaking of elements x_i and x_j all the time. Don't name the elements of X unless necessary. Then you can refer to elements x and y of X in your subsequent discussion, without needing subscripts; or you can refer to x_1 and x_2 as specified elements of X.

16. Display important formulas on a line by themselves. If you need to refer to some of these formulas from remote parts of the text, give reference numbers to all of the most important ones, even if they aren't referenced.

17. Sentences should be readable from left to right without ambiguity. Bad examples: "Smith remarked in a paper about the scarcity of data." "In the theory of rings, groups and other algebraic structures are treated."

18. Small numbers should be spelled out when used as adjectives, but not when used as names (i.e., when talking about numbers as numbers).

    > Bad: The method requires 2 passes.
    > Good: Method 2 is illustrated in Fig. 1; it requires 17 passes. The count was increased by 2. The leftmost 2 in the sequence was changed to a 1.

19. Capitalize names like Theorem 1, Lemma 2, Algorithm 3, Method 4.

20. Some handy maxims:
    > Watch out for prepositions that sentences end with.
    > When dangling, consider your participles.
    > About them sentence fragments.
    > Make each pronoun agree with their antecedent.
    > Don't use commas, which aren't necessary.
    > Try to never split infinitives.

21. Some words frequently misspelled by computer scientists:

implement	not	impliment
complement	not	compliment
occurrence	not	occurence
dependent	not	dependant
auxiliary	not	auxillary
feasible	not	feasable
preceding	not	preceeding
referring	not	refering
category	not	catagory
consistent	not	consistant
PL/I	not	PL/1
descendant (noun)	not	descendent
its (belonging to it)	not	it's (it is)

    The following words are no longer being hyphenated in current literature:
    > nonnegative
    > nonzero

22. Don't say "which" when "that" sounds better. The general rule nowadays is to use "which" only when it is preceded by a comma or by a preposition, or when it is used interrogatively. Experiment to find out which is better, "which" or "that", and you'll understand this rule.

    > Bad: Don't use commas which aren't necessary.
    > Good: Don't use commas that aren't necessary.

    Another common error is to say "less" when it should be "fewer".

23. In the example at the bottom of §2 below, note that the text preceding displayed equations (1) and (2) does not use any special punctuation. Many people would have written

    > ... of "nonincreasing" vectors:

    $$A_n = \{(a_1, \ldots, a_n) \in N^n \mid a_1 \geq \cdots \geq a_n\}. \qquad (1)$$

    > If C and P are subsets of N^n, let:

    $$L(C, P) = \ldots$$

    and those colons are wrong.

24. The opening paragraph should be your best paragraph, and its first sentence should be your best sentence. If a paper starts badly, the reader will wince and be resigned to a difficult job of fighting with your prose. Conversely, if the beginning flows smoothly, the reader will be hooked and won't notice occasional lapses in the later parts.

    Probably the worst way to start is with a sentence of the form "An x is y." For example,

    > Bad: An important method for internal sorting is quicksort.
    > Good: Quicksort is an important method for internal sorting, because ...
    > Bad: A commonly used data structure is the priority queue.
    > Good: Priority queues are significant components of the data structures needed for many different applications.

25. The normal style rules for English say that commas and periods should be placed inside quotation marks, but other punctuation (like colons, semicolons, question marks, exclamation marks) stay outside the quotation marks unless they are part of the quotation. It is generally best to go along with this illogical convention about commas and periods, because it is so well established, except when you are using quotation marks to describe some text as a specific string of symbols. For example,

    > Good: Always end your program with the word "end".

    On the other hand, punctuation should always be strictly logical with respect to parentheses and brackets. Put a period inside parentheses if and only if the sentence ending with that period is entirely within the parentheses. The punctuation within parentheses should be correct, independently of the outside context, and the punctuation outside the parentheses should be correct if the parenthesized statement would be removed.

    > Bad: This is bad, (although intentionally so.)

26. Resist the temptation to use long strings of nouns as adjectives: consider the packet switched data communication network protocol problem.

    In general, don't use jargon unnecessarily. Even specialists in a field get more pleasure from papers that use a nonspecialist's vocabulary.

    > Bad: "If $\mathbf{L}^+(P, N_0)$ is the set of functions $f \colon P \to N_0$ with the property that
    >
    > $$\underset{n_0 \in N_0}{\exists} \underset{p \in P}{\forall} p \geq n_0 \Rightarrow f(p) = 0$$
    >
    > then there exists a bijection $N_1 \to \mathbf{L}^+(P, N_0)$ such that if $n \to f$ then
    >
    > $$n = \prod_{p \in P} p^{f(p)}.$$
    >
    > Here P is the prime numbers and $N_1 = N_0 \sim \{0\}$."

Better: "According to the 'fundamental theorem of arithmetic' (proved in ex. 1.2.4–21), each positive integer u can be expressed in the form

$$u = 2^{u_2} 3^{u_3} 5^{u_5} 7^{u_7} 11^{u_{11}} \ldots = \prod_{p \text{ prime}} p^{u_p},$$

where the exponents $u_2, u_3, \ldots$ are uniquely determined nonnegative integers, and where all but a finite number of the exponents are zero."

[The first quotation is from Carl Linderholm's neat satirical book *Mathematics Made Difficult*; the second is from D. Knuth's *Seminumerical Algorithms*, Section 4.5.2.]

27. When in doubt, read *The Art of Computer Programming* for outstanding examples of good style.

[That was a joke. Humor is best used in technical writing when readers can understand the joke only when they also understand a technical point that is being made. Here is another example from Linderholm:

"... $\emptyset D = \emptyset$ and $N\emptyset = N$, which we may express by saying that $\emptyset$ is absorbing on the left and neutral on the right, like British toilet paper."

Try to restrict yourself to jokes that will not seem silly on second or third reading. And don't overuse exclamation points!]

§2. An Exercise on Technical Writing

In the following excerpt from a term paper, N denotes the nonnegative integers, N^n denotes the set of n-tuples of nonnegative integers, and $A_n = \{(a_1, \ldots, a_n) \in N^n \mid a_1 \geq \cdots \geq a_n\}$. If $C, P \subset N^n$, then $L(C, P)$ is defined to be $\{c + p_1 + \cdots + p_m \mid c \in C, m \geq 0,$ and $p_j \in P$ for $1 \leq j \leq m\}$. We want to prove that $L(C, P) \subseteq A_n$ implies $C, P \subseteq A_n$.

The following proof, directly quoted from a sophomore term paper, is mathematically correct (except for a minor slip) but stylistically atrocious:

$L(C, P) \subset A_n$
$C \subset L \Rightarrow C \subset A_n$
Spse $p \in P$, $p \notin A_n \Rightarrow p_i < p_j$ for $i < j$
$c + p \in L \subset A_n$
$\therefore c_i + p_i \geq c_j + p_j$ but $c_i \geq c_j \geq 0, p_j \geq p_i \therefore (c_i - c_j) \geq (p_j - p_i)$
but $\exists$ a constant $k \ni c + kp \notin A_n$
let $k = (c_i - c_j) + 1 \quad c + kp \in L \subset A_n$
$\therefore c_i + kp_i \geq c_j + kp_j \Rightarrow (c_i - c_j) \geq k(p_j - p_i)$
$\Rightarrow k - 1 \geq k \cdot m \quad k, m \geq 1 \quad$ Contradiction
$\therefore p \in A_n$
$\therefore L(C, P) \subset A_n \Rightarrow C, P \subset A_n$ and the lemma is true.

A possible way to improve the quality of the writing:

Let N denote the set of nonnegative integers, and let

$$N^n = \{(b_1, \ldots, b_n) \mid b_i \in N \text{ for } 1 \leq i \leq n\}$$

be the set of n-dimensional vectors with nonnegative integer components. We shall be especially interested in the subset of "nonincreasing" vectors,

$$A_n = \{(a_1, \ldots, a_n) \in N^n \mid a_1 \geq \cdots \geq a_n\}. \tag{1}$$

If C and P are subsets of N^n, let

$$L(C, P) = \{c + p_1 + \cdots + p_m \mid c \in C, m \geq 0, \text{ and } p_j \in P \text{ for } 1 \leq j \leq m\} \tag{2}$$

be the smallest subset of N^n that contains C and is closed under the addition of elements of P. Since A_n is closed under addition, $L(C, P)$ will be a subset of A_n whenever C and P are both contained in A_n. We can also prove the converse of this statement.

Lemma 1. *If $L(C, P) \subseteq A_n$ and $C \neq \emptyset$, then $C \subseteq A_n$ and $P \subseteq A_n$.*

Proof. (Now it's your turn to write it up beautifully.)

§3. An Answer

Here is one way to complete the exercise in the previous section. (But please try to WORK IT YOURSELF BEFORE READING THIS.) Note that a few clauses have been inserted to help keep the reader synchronized with the current goals and subgoals and strategies of the proof. Furthermore the notation $(b_1, \ldots, b_n)$ is used instead of $(p_1, \ldots, p_n)$, in the second paragraph below, to avoid confusion with formula (2).

Proof. Assume that $L(C, P) \subseteq A_n$. Since C is always contained in $L(C, P)$, we must have $C \subseteq A_n$; therefore only the condition $P \subseteq A_n$ needs to be verified.

If P is not contained in A_n, there must be a vector $(b_1, \ldots, b_n) \in P$ such that $b_i < b_j$ for some $i < j$. We want to show that this leads to a contradiction.

Since the set C is nonempty, it contains some element $(c_1, \ldots, c_n)$. We know that the components of this vector satisfy $c_1 \geq \cdots \geq c_n$, because $C \subseteq A_n$.

Now $(c_1, \ldots, c_n) + k(b_1, \ldots, b_n)$ is an element of $L(C, P)$ for all $k \geq 0$, and by hypothesis it must therefore be an element of A_n. But if we take $k = c_i - c_j + 1$, we have $k \geq 1$ and
$$c_i + kb_i \geq c_j + kb_j,$$
hence
$$c_i - c_j \geq k(b_i - b_j). \tag{3}$$
This is impossible, since $c_i - c_j = k - 1$ is less than k, yet $b_j - b_i \geq 1$. It follows that $(b_1, \ldots, b_n)$ must be an element of A_n. ∎

Note that the hypothesis $C \neq \emptyset$ is necessary in Lemma 1, for if C is empty the set $L(C, P)$ is also empty regardless of P.

[This was the "minor slip."]

BUT ... don't always use the first idea you think of. The proof above actually commits another sin against mathematical exposition, namely the unnecessary use of proof by contradiction. It would have been better to use a direct proof:

Let $(b_1, \ldots, b_n)$ be an arbitrary element of P, and let i and j be fixed subscripts with $i < j$; we wish to prove that $b_i \geq b_j$. Since C is nonempty, it contains some element $(c_1, \ldots, c_n)$. Now the vector $(c_1, \ldots, c_n) + k(b_1, \ldots, b_n)$ is an element of $L(C, P)$ for all $k \geq 0$, and by hypothesis it must therefore be an element of A_n. But this means that $c_i + kb_i \geq c_j + kb_j$, i.e.,
$$c_i - c_j \geq k(b_j - b_i), \tag{3}$$
for arbitrarily large k. Consequently $b_j - b_i$ must be zero or negative.

We have proved that $b_j - b_i \leq 0$ for all $i < j$, so the vector $(b_1, \ldots, b_n)$ must be an element of A_n. ∎

This form of the proof has other virtues too: It doesn't assume that the b_i's are integer-valued, and it doesn't require stating that $c_1 \geq \cdots \geq c_n$.

§4. Excerpts from class, October 7 [notes by TLL]

Our first serious business involved examining "the worst abusers of the 'Don't use symbols in titles' rule." Professor Knuth (hereafter known as Knuth) displayed a paper by Gauss that had a long displayed formula in the title. He showed us a bibliography he's preparing that references not only that paper but another with even more symbols in the title. (Such titles make more than bibliographies difficult; they make bibliographic data retrieval systems and keyword-in-context produce all sorts of hiccups.)

In his bibliography Knuth has tried to keep his citations true to the original sources. The bibliography contains mathematical formulas, full name spellings (even alternate spellings when common), and completely spelled-out source journal names. (This last may be unusual enough that some members of a field may be surprised to see the full journal name written out, but it's a big help to novices who want to find it in the library.)

We spent the rest of class going over some of solutions that students had turned in for the exercise of §2 (each sample anonymous). He cautioned us that while he was generally pleased by the assignments, he was going to be pointing out things that could be improved. The following points were all made in the process of going through these samples.

> In certain instances, people did not understand what constitutes a proof. Fluency in mathematics is important for Computer Science students but will not be taught in this class.
>
> Not all formulas are equations. Depending on the formula, the terms 'relation', 'definition', 'statement', or 'theorem' might be used.
>
> Computer Scientists must be careful to distinguish between mathematical notation and programming language notation. While it may be appropriate to use $p[r]$ in a program, in a formal paper it is probably better to use p with a subscript of r. As another example, it is not appropriate to use a star ($*$) to denote multiplication in a paper about mathematics. Just say xy.
>
> Some people called p an element of P and p_r an element of p. Everything was an "element." It's better to call p_r a "component" of p, thus distinguishing two kinds of subsidiary relationships.
>
> It is natural in mathematics to hold off some aspects of your definition — to "place action before definition" (as in '$p(x) < p(y)$ for some $x < y$'). But it is possible to carry this too far, if too much is being held back. The best location for certain definitions is a subjective matter.
>
> Remember to put words between adjacent formulas.
>
> When you use ellipses, such as $(P_1, \ldots, P_n)$, remember to put commas before and after the three dots. When placing ellipses between commas the three dots belong on the same level as the commas, but when the ellipsis is bracketed by symbols such as '+' or '<' the dots should be at mid-level.

Be careful with the spacing around ellipses. The character string '...;' looks strange (it should have more space after the last dot). All kinds of accidents happen concerning spaces in formulas. Typesetting systems like TeX have built-in rules to cover 99% of the cases, but if you write a lot of mathematics you will get bitten.

Linebreaks in the middle of formulas are undesirable. There are ways to enforce this with TeX (as well as other text formatting systems). People who use TeX and wish to use the vertical bar and the empty set symbol in notation like '$\{c \mid c \in \emptyset\}$' should be aware of the TeX commands `\mid` and `\emptyset`.

Comments such as, "We demonstrate the second conclusion by contradiction," and "There must be a witness to the unsortedness of P," are useful because they tell the reader what is going on or bring in new and helpful vocabulary.

Numbering all displayed formulas is usually a bad idea; number the important ones only. Extraneous parentheses can also be distracting. For example, in the phrase "let k be $(c_i - c_j) + 1$," the parentheses should omitted.

You can overdo the use of any good tool. For instance, you could overuse typographic tools by having 20 different fonts in one paper.

Two more topics were touched on (and are sure to be discussed further): the use of 'I' in technical writing, and the use of past or present tense in technical writing.

Knuth says that Mary-Claire van Leunen defends the use of 'I' in scholarly articles, but that he disagrees (unless the identity of the author is important to the reader). Knuth likes the "teamwork" aspect of using 'we' to represent the author and reader together. If there are multiple authors, they can either "revel in the ambiguity" of continuing to use 'we', or they can use added disambiguating text. If one author needs to be mentioned separately, the text can say 'one of the authors (DEK)', or 'the first author', but not 'the senior author'.

Knuth (hereafter known as Don) recommends that one of two approaches be used with respect to tenses of verbs: Either use present tense throughout the entire paper, or write sequentially. Sequential writing means that you say things like, "We saw this before. We will see this later." The sequential approach is more appropriate for lengthy papers. You can use it even more effectively by using words of duration: "We observed this long ago. We saw the other thing recently. We will prove something else soon."

§5. Excerpts from class, October 9 [notes by TLL]

"I'm thinking about running a contest for the best Pascal program that is also a sonnet," was the one of the first sentences out of Don's mouth on the topic of the exact definition of "Mathematical Writing." He admitted that such a contest was "probably not the right topic for this course." However, a program (presumably even an iambic pentameter program) is among the documents that he will accept as the course term paper. He will accept articles for professional journals, chapters of books or theses, term papers for other courses, computer programs, user manuals or parts thereof: anything that falls into a definite genre where you have a specific audience in mind and the technical aspect is significant.

We spent the rest of class continuing to examine the homework assignment. In the interest of succinct notes, I have replaced many literal phrases by their generic equivalents. For example, I might have replaced '$A > B$' by '⟨relation⟩'. This time I have divided the comments into two sets: those dealing with what I will call "form" (parentheses, capitalization, fonts, etc.) and those dealing with "content" (wording, sentence construction, tense, etc.).

First, the comments concerning form:

> Don't overdo the use of colons. While the colon in 'Define it as follows:' is fine, the one in 'We have: ⟨formula⟩' should be omitted since the formula just completes the sentence. Some papers had more colons than periods.

> Should the first word after a colon be capitalized? Yes, if the phrase following the colon is a full sentence; No, if it is a sentence fragment. (This is not "yet" a standard rule, but Don has been trying it for several years and he likes it.)

> While too many commas will interfere with the smooth flow of a sentence, too few can make a sentence difficult to read. As examples, a sentence beginning with 'Therefore, ' does not need the comma following 'therefore'. But 'Observe that if ⟨symbol⟩ is ⟨formula⟩ then so is ⟨symbol⟩ because ⟨reasoning⟩' at least needs a comma before 'because'.

> Putting too many things in parentheses is a stylistic thing that can get very tiring. (When Don moves from his original, handwritten draft to a typed, computer-stored version his most frequent change is to remove extra parentheses.)

> Among the parentheses most in need of removal are nested parentheses. To this end, it is better to write '(Definition 2)' than '(definition (2))'. Unfortunately, however, you can't use the former if the definition was given in displayed formula (2). Then it's probably best to think of a way to avoid the outer parentheses altogether.

> In some cases your audience may expect nested parentheses. In this case (or in any other case when you feel you must have them), should the outer pair be changed to brackets (or curly-braces)? This was once the prevailing convention, but it is now not only obsolete but potentially dangerous; brackets and curly braces have semantic content for many scientific professionals. ("The world is short of delimiters," says

Don.) Typographers help by using slightly larger parentheses for the outer pair in a nested set.

An entire paper or proof in capital letters is distracting. It gives the impression of sustained shouting. Same goes for boldface, etc.

Paul Halmos introduced the handy convention of placing a box at the end of a proof; this box serves the same function as the initials 'Q.E.D.'. If you use such a box, it seems best to leave a space between it and the final period.

Try to make it clear where new paragraphs begin. When using displayed formulas, this can become confusing unless you are careful.

Using notational or typographic conventions can be helpful to your readers (as long as your convention is appropriate to your audience). Boldface symbols or arrows over your vectors are each appropriate in the correct context. When using a raised 'st' in phrases such as 'the $i+1^{st}$ component', it's better to use roman type: '$i+1^{\text{st}}$'. Then it's clear that you aren't speaking of "1 raised to the power st."

Avoid "psychologically bad" line breaks. This is subjective, but you can catch many such awkward breaks by not letting the final symbol lie on a line separate from the rest of its sentence. If you are using TeX, a tilde (~) in place of a space will cause the two symbols on either side of the tilde to be tied together. (Other text processors also have methods to disallow line breaks at specific points.)

Some of us are much better at spelling than others of us. Those of us who are not naturally wonderful spellers should learn to use spelling checkers.

Allowing formulas to get so long that they do not format well or are unnecessarily confusing "violates the principle of 'name and conquer' that makes mathematics readable." For example, '$v = c + u(c_i - c_j + 1)$' should be '$v = c + ku$, where $k = c_i - c_j + 1$', if you're going to do a lot of formula manipulation in which $(c_i - c_j + 1)$ remans as a unit.

Be stingy with your quotation marks. "Three cute things in quotes is a little too cute."

Remember to minimize subscripts. For example, 'p_i is an element of P' could more easily be 'p is an element of P'.

Remember to capitalize words like theorem and lemma in titles like Lemma 1 and Theorem 23.

Remember to place words between adjacent formulas. A particularly bad example was, "Add p k times to c."

Be careful to define symbols before you use them (or at least to define them very near where you use them).

Don't get hung up on one or two styles of sentences. The following sort of thing can become very monotonous:
> Thus, $----$.
> Consequently, $----$.
> Therefore, $----$.
> And so, $----$.

On the other hand, parallelism should be used when it is the point of the sentence.

Now the comments involving content:

Try to make sentences easily comprehensible from left to right. For example, "We prove that ⟨grunt⟩ and ⟨snort⟩ implies ⟨blah⟩." It would be better to write "We prove that the two conditions ⟨grunt⟩ and ⟨snort⟩ imply ⟨blah⟩." Otherwise it seems at first that ⟨grunt⟩ and ⟨snort⟩ are being proved.

While guidelines have been given for the use of the word 'that', the final placement must be dictated by cadence and clarity. Read your words aloud to yourself.

The word 'shall' seems to be a natural word for definitions to many mathematical readers, but it is considered formal by younger members of the audience.

Be precise in your wording. If you mean "not nonincreasing," don't say "increasing"; the former means that $p_j < p_{j+1}$ for *some* j, while the latter that $p_j < p_{j+1}$ for *all* j.

Mixed tenses on the same subject are awkward. For example, "We assume now ⟨grunt⟩, hoping to show a contradiction," is better than, "We assume now ⟨grunt⟩, and will show that this leads to a contradiction."

Many people used the ungainly phrase "Assume by contradiction that ⟨blah⟩." It is better to say, "The proof that ⟨blah⟩ is by contradiction," and even better to say "To prove ⟨grunt⟩, let us assume the opposite and see what happens."

In general, a conversational tone giving signposts and clearly written transition paragraphs provides for pleasant reading. One especially easy-to-read proof contained the phrases "The operative word is zero," "The lemma is half proved," and "We divide the proof into two parts, first proving ⟨blah⟩ and then proving ⟨grunt⟩."

You can give relations in two ways, either saying '$p_i < p_j$' or '$p_j > p_i$'. The latter is for "people who are into dominance," Don says, but the former is much easier for a reader to visualize after you've just said '$p = (p_1, p_2, \ldots, p_n)$ and $i < j$'. Similarly, don't say '$i < j$ and $p_j < p_i$'; keep i and j in the same relative position.

§6. Excerpts from class, October 12 [notes by TLL]

Don opened class by saying that up until now he has been criticizing our writing; now he will show us what he does to his own. Perhaps apropos showing us his own writing he quoted Dijkstra: "A good teacher will teach his students the importance of style and how to develop their own style—not how to mimic his."

First he showed us a letter from Bob Floyd. The letter opened by saying 'Don, Please stop using so many exclamation points!' and closed with at least five exclamation points. After receiving this letter he looked in *The Art of Computer Programming* and found about two exclamation points per page. (Among the other biographical tidbits we learned at this class were that Don went to secretarial school, types 80 words per minute, and once knew two kinds of shorthand.)

Don is writing a book with Oren Patashnik and Ron Graham. The book is entitled *Concrete Mathematics* and is to be used for CS 260. He showed us two copies of Chapter Five of this book: one copy he called "Before" and one he called "After".

The Before copy actually came into existence long after the work on the book began. Oren wrote several drafts using the LaTeX book style, and then the authors availed themselves of the services of a book designer. The designer decided how wide the text was, what fonts were to be used, what chapter headings looked like, and a host of other things. The designer, at the authors request, has left room in the inner margins for "graffiti". That is, for informal snappy comments from the peanut gallery. (This idea was "stolen" from the booklet *Approaching Stanford*.)

The After copy is just another formally typeset revision of the Before copy. Neither copy has yet been through a professional copy editor. Having now mentioned copy editors and book designers, Don said, "In these days of author self-publishing, we must not forget the value of professionals." (Actually, the copy editor was first mentioned when an error in punctuation was displayed on the screen.)

Upon receiving a question from the audience concerning how many times he actually rewrites something, Don told us (part of) his usual rewrite sequence:

His first copy is written in pencil. Some people compose at a terminal, but Don says, "The speed at which I write by hand is almost perfectly synchronized with the speed at which I think. I type faster than I think so I have to stop, and that interrupts the flow."

In the process of typing his handwritten copy into the computer he edits his composition for flow, so that it will read well at normal reading speed. Somewhere around here the text gets TeXed, but the description of this stage was tangled up with the description of the process of rewriting the composition. Of course, rewriting does not all occur at any one stage. As Don said, "You see things in different ways on the different passes. Some things look good in longhand but not in type."

While discussing his own revisions, he mentioned those of two other Computer Science authors. Nils Nilsson had at least five different formal drafts of his "Non-Monotonic Reasoning" chapter. Tony Hoare revised the algorithm in his paper on "Communicating Sequential Processes" more than a dozen times over the course of two years.

Don, obviously a fan of rewriting in general, told us that he knows of many computer programs that were improved by scrapping everything after six months and starting from scratch. He said that this exact approach was used at Burroughs on their Algol compiler in 1960 and the result was what Don considers to be one of the best computer programs he has ever seen. On the limits of the usefulness of rewriting, he did say, "Any writing can be improved. But eventually you have to put something out the door."

The last part of class was spent discussing the font used in the coming book: Euler. The Euler typeface was designed by Hermann Zapf ("probably the greatest living type designer") and is an especially appropriate font to use in a book that is all about Euler's work. The idea of the face is to look a bit handwritten. For example, the zero to be used for mathematics has a point at the apex because "when people write zeros, they never really close them". This zero is different from the zero used in the text (for example, in a date), so book preparation with Euler needs more care than usual. You have to distinguish mathematical numerals from English-language numerals in the manuscript.

Somebody asked about 'all' versus 'all of'. Which should it be? Answer: That's a very good question. Sometimes one way sounds best, sometimes the other. You have to use your ear. Another tricky business is the position of 'only' and 'also'; Don says he keeps shifting those words around when he edits for flow.

§7. Excerpts from class, October 14 [notes by PMR]

Don discussed the labours of the book designer and showed us specimen "page plans" and example pages. The former are templates for the page and show the exact dimensions of margins, paragraphs, etc. His designer also suggested a novel scheme for equations: They are to be indented much like paragraphs rather than being centered in the traditional way. We also saw conventions for the display of algorithms and tables. Although Don is doing his own typesetting, he is using the services of the designer and copy editor. These professionals are well worth their keep, he said. Economists in the audience were not surprised to hear that the prices of books bear almost no relation to their production costs. Hardbacks are sometimes cheaper to produce than paperbacks. For those interested in such things, Don recommended a paperback entitled *One Book / Five Ways* (available in the Bookstore) that describes the entire production process by means of actual documents.

Returning to the editing of his Concrete Maths text, Don went through more of the Before and After pages he began to show us on Monday, picking out specific examples that illustrate points of general interest.

He exhorted writers to try to put themselves in their readers' shoes: "Ask yourself what the reader knows and expects to see next at some point in the text." Ideally, the finished version reads so simply and smoothly that one would never suspect that had been rewritten at all. For example, part of the Concrete Math draft said

> (Before) The general rule is (...) and it is particularly valuable because (...). The transformation in (5.12) is called (...). It is easily proved since (... and ...).

Reading this at speed and in context made it clear that readers would be hanging on their chairs wondering why the rule was true; so we should first tell them why, before stressing the rule's significance:

> (After) The general rule is (...) and it is easily proved since (... and ...).
>
> [new paragraph] Identity (5.12) is particularly valuable because (...). It is called (...).

Don's favorite dictionary was of no help on the question of 'replace with' vs. 'replace by'. The phrase 'by replacing $--$ by $--$' is bad (due to the repetition), and 'by replacing $--$ with $--$' seems worse. In this case the solution is to choose another word: 'by changing $--$ to $--$'.

As a very general rule, try reading at speed. You will often get a much better sense of the rhythm of the sentence than you did when you wrote it.

It is a bad idea to display false equations. The reader's eye is apt to alight upon them in the text and treat them as gospel. It is much better to put them into the text, as in "So the equation ' ... ' is always false!"

Be sure that the antecedent of any pronoun that you use is clear. For example, the previous paragraph has two sentences beginning 'It is ... '; they are fine. But sometimes such a sentence structure is troublesome because 'it' seems to be referring to an object under discussion. For example,

> (Before) Two things about the derivation are worthy of note. First, it's a great convenience to be summing ... .
>
> (After) Two things about this derivation are worthy of note. First, we see again the great convenience of summing ... .

Towards the end of the editing process you will need to ensure that you don't have a page break in the middle of a displayed formula. Often you'll simply have to think up something else to say to fill up the page, thus pushing the displayed formula entirely onto the next page. Try to think of this as a stimulus to research!

Let proofs follow the same order as definitions, e.g., where you have to deal with several separate cases.

Hyphens, dashes, and minus signs are distinct and should not be used interchangeably. The shortest is the hyphen. The next is the en-dash, as in 'lines 10–18'. Longer still is the minus sign, used in formulæ: '$10 - 18 = -8$'. The longest of all is the em-dash—used in sentences.

When proofreading you may catch technical errors as well as stylistic errors. Think about the mathematics too, not just the prose. For example, the book was discussing a purported argument that 0^0 should be undefined "because the functions x^0 and 0^x have different limiting values when $x \to 0$". Don revised this statement to "... when x decreases to 0," because 0^x is undefined when $x \to 0$ through negative values.

When you use the word 'instead', be clear about the contrast you are drawing. The reader should immediately understand what you are referring to:

(Before) And when $x = -1$ instead, ...

(After) And when $x = -1$ instead of $+1$, ...

Notice the helpful use of a redundant '+' sign here.

Use the present tense for timeless facts. Things that we proved some time ago are nevertheless still true.

Try to avoid repeating words in a sentence.

(Before) -- approach the values -- fill in the values -- .

(After) -- approach the values -- fill in the entries -- .

In answer to a question from the class, Don suggested giving page numbers only for remote references (to equations, say). Usually it is enough to say 'using Equation 5.14' or whatever. It becomes unwieldy to give page numbers for every single such reference. A member of the class suggested the 'freeway method' for numbering tables; number them with the page number on which they appear. Don confessed that he hadn't thought of this one. Sounds like a neat idea.

The formula

(Before) $$\sum_{k \leq m} \binom{r}{k} \left(k - \frac{r}{2}\right) = -\frac{m+1}{2} \binom{r}{m+1}$$

looks a bit confusing because of the minus sign on the right, so Don changed it to

(After) $$\sum_{k \leq m} \binom{r}{k} \left(\frac{r}{2} - k\right) = \frac{m+1}{2} \binom{r}{m+1}.$$

There may be many ways to write a formula; you have the freedom to select the best. (This change also propagated into the subsequent text, where a reference to 'the factor $(k - r/2)$' had to be changed to 'the factor $(r/2 - k)$'.

Somebody saw an integral sign on that page and asked about the relative merits of

$$\int_{-\infty}^{a} f(x)\,dx$$

versus other notations like

$$\int\limits_{-\infty}^{a} f(x)\,dx \qquad \int\limits_{x=-\infty}^{x=a} f(x)\,dx.$$

Don said that putting limits above and below, instead of at the right, traded vertical space for horizontal space, so it depends on how wide your formulas are. Both forms are used.

Whichever form you adopt should be consistent throughout an entire book. Somebody suggested
$$\int_{x=-\infty}^{x=a} f(x)\,dx$$
but Don pooh-poohed this.

He said that major writing projects each have their own style; you get to understand the style that's appropriate as you write more and more of the book, just as novelists learn about the characters they are creating as they develop a story. In *Concrete Mathematics* he is learning how to use the idea of "graffiti in the margin" as he writes more. One nice application is to quote from the first publications of important discoveries; thus famous mathematicians like Leibniz join the writers of 20th century graffiti.

§8. Excerpts from class, October 16 [notes by TLL]

We continued to examine Before and After pages from the book of which Don is a co-author. The following points were made in reference to changes Don decided to make.

> When long formulas don't fit, try to break the lines logically. In some cases the author can even change some of the math (perhaps by introducing a new symbol) to make the formula placement less jarring. Such a change is best made by the author, since the choice of how to display a complex expression is an important part of any mathematical exposition.
>
> Sometimes moving a formula from embedded text to one separately displayed will allow the formula to be more logically divided. The placement of the equals sign ($=$) is different for line breaks in the middle of displayed versus embedded formulas: The break comes after the equals sign in an embedded formula, but before the equals sign in a display.
>
> While editing for flow, sentences can be broken up by changing semicolons to periods; or if you want the sentences to join into a quickly moving blur, you can change periods to semicolons. Breaking existing paragraphs into smaller paragraphs can also be helpful here.
>
> While making such changes make sure to preserve clarity. For example, make sure that any sentences you create that begin with conjunctions are constructed clearly, and that words like 'it' have clear antecedents. (Sentences that begin with the word 'And' are not always evil.)
>
> Make sure your variable names are not misleading. Variable names that are too similar to conceptually unrelated variables can be confusing. Systematic variable renaming is one of the advantages of text editors.
>
> We noted last time that present tense is correct for facts that are still true; but it is okay to use past tense for "facts" that have turned out to be in error.

One of the most common errors that mathematicians make when they get their own typesetting systems is to over-use the form of fraction with a horizontal bar $\left(\frac{1+x}{y}\right)$ rather than a slash $\left((1+x)/y\right)$. The stacked form can lead to tiny little numbers—especially when they are used in exponents. One of the most common changes that mathematical copy editors make is to slash a mathematician's fractions. (They even know that they have to add parentheses when they do this.)

Exercises are some of the most difficult parts of a book to write. Since an exercise has very little context, ambiguity can be especially deadly; a bit of carefully chosen redundancy can be especially important. For this reason, exercises are also the hardest technical writing to translate to other languages.

Copyright law has changed, making it technically necessary to give credit to all previously published exercises. Don says that crediting sources is probably sufficient (he doesn't plan to write every person referenced in the exercises for his new book, unless the publisher insists). Tracing the history of even well-known theorems can be difficult, because mathematicians have tended to omit citations. He recently spent four hours looking through the collected works of Lagrange trying to find the source of "Lagrange's inequality," but he was unsuccessful. Considering the benefit to future authors and readers, he's not too unhappy with the new law.

We can dispense with some of our rhetorical guidelines when writing the answers to exercises. Answers that are quick and pithy, and answers that start with a symbol, are quite acceptable.

§9. Excerpts from class, October 16 (continued) [notes by TLL]

From esoteric mathematics we moved on to reference books. Don showed us six such books that he likes to have next to him when he writes. [And he added a seventh later.]

1. *The Oxford English Dictionary* (usually called the OED). He showed us the two volume "squint print" edition rather than the 16-volume set. This compact edition is often offered as a bonus given to new members upon joining a book club. (There is a project in Toronto that will soon have the entire OED online.)

2. The *OED Supplement*. The supplement brings the OED up to date. The supplement comes in four volumes, each of which costs $100 or more, so you may have to go to the library for this one.

3. *The American Heritage Dictionary*. Don likes this dictionary because of the usage notes and the Appendix containing Indo-European root words. (For example, the usage notes will help you choose between 'compare to' or 'compare with' in a specific sentence.)

4. *The Longman Dictionary of Contemporary English*. Instead of the historical words found in the previously mentioned dictionaries, this one has the words used on the street. Current slang and popular usage are explained in very simple English.

(For example, the nuances of 'mind-bending' versus 'mind-blowing' versus 'mind-boggling' are explained.)

5. *Webster's New Word Speller Divider.* Don said that people who don't spell well find this book to be quite useful. [I saw no indication that *he* actually uses it, though.]

6. *Roget's Thesaurus.* This book is a synonym dictionary. Don says that he owns two, one for home and one for his Stanford office, and he uses them in many different ways: when he knows that a word exists but has forgotten it; when he wants to avoid repetition; when he wants to define a new technical term or a new name for a paper or program.

7. *Webster's Dictionary of English Usage.* A wonderful new (1989) resource, which goes well beyond the *American Heritage* usage notes. It's filled with choice examples and is a joy to read.

The issue of British versus American dialect came up. When writing for international audiences, should we use British or American spellings and conventions? Don says he agrees with the rule that Americans should write with their own spellings and the British should do the same. The two styles should be mixed only when, say, an American writes about the 'labor of the British Labour Party'. (Readers of these classnotes will now understand why TLL and PMR spell some words differently.)

§10. Excerpts from class, October 19   [notes by TLL]

Should this course have been named "Computer Scientifical Writing" or "Informatical Writing" rather than "Mathematical Writing"? The Computer Science Department is offering this class, but until now we have been talking about topics that are generally of concern to all writers who use mathematics. Today we begin to discuss topics specific to the writing of Computer Science.

We are not abandoning mathematical concerns; Don says that a technical typist in Computer Science must know all that a Math department typist must know plus quite a bit more. He showed us two examples where mathematical journals had trouble presenting programs, algorithms, or concrete mathematics in papers he wrote. In order to solve the first problem, Don had to convince the typesetters at *Acta Arithmetica* to create "floor" and "ceiling" functions by carving off small pieces of the metal type for square brackets. The second problem had to do with typographic conventions for computer programs; *The American Mathematical Monthly* was using different fonts for the same symbol at different points in a procedure, was interchangeably using ":=", ": =", and "=:" to represent an assignment symbol.

Stylistic conventions for programming languages originated with Algol 60. Prior to 1960, FORTRAN and assembly languages were displayed using all uppercase letters in variable width fonts that did not mix letters and numbers in a pleasant manner. Fortunately, Algol's visual presentation was treated with more care: Myrtle Kellington of ACM worked from the beginning with Peter Naur (editor of the Algol report) to produce a set of conventions concerning, among other things, indentation and the treatment of reserved words.

Don found the prevailing variable-width fonts unacceptable for use in the displayed computer programs in Volume 1 of *The Art of Computer Programming*, and he insisted that he needed fixed width type. The publishers initially said that it wasn't possible, but they eventually found a way to mix `typewriter` style with roman, **bold**, and *italic*.

Don says he had a difficult time trying to decide how to present algorithms. He could have used a specific programming language, but he was afraid that such a choice would alienate people (either because they hated the language or because they had no access to the language). So he decided to write his algorithms in English.

His Algorithms are presented rather like Theorems with labeled steps; often they have accompanying (but very high-level) flow charts (a technique he first saw in Russian literature of the 1950s). The numbered steps have parenthetical remarks that we would call comments; after 1968 these parenthetical remarks are often invariant relations that can be used in a formal proof of program correctness.

Don has received many letters complimenting him on his approach, but he says it is not really successful. Explaining why, he said, "People keep saying, 'I'm going to present an algorithm in Knuth's style,' and then they completely botch it by ignoring the conventions I think are most important. This style must just be a personal style that works for me. So get a personal style that works for you." In recent papers he has used the pidgin Algol style introduced by Aho, Hopcroft, and Ullman; but he will not change his style for the yet-unfinished volumes of *The Art of Computer Programming* because he wants to keep the entire series consistent.

Don says that a computer program is a piece of literature. ("I look forward to the day when a Pulitzer Prize will be given for the best computer program of the year.") He says that, apart from the benefit to be gained for the readers of our programs, he finds that treating programs in this manner actually helps to make them run smoothly on the computer. ("Because you get it right when you have to think about it that way.")

He gave us a reprint of "Programming Pearls" by Jon Bentley, from *Communications of the ACM* **29** (May 1986), pages 364–369, and told us we had best read it by Wednesday since it will be an important topic of discussion. Don, who was 'guest oyster' for this installment of "Programming Pearls," warned us that "this represents the best thing to come out of the TeX project. If you don't like it, try to conceal your opinions until this course is over."

Bentley published that article only after Don had first published the idea of "literate programming" in the British *Computer Journal*. (Don says that he chose the term in hopes of making the originators of the term "structured programming" feel as guilty when they write illiterate programs as he is made to feel when he writes unstructured programs.) When Bentley wanted to know why Don did not publish this in America, Don said that Americans are illiterate and wouldn't care anyway. Bentley seems to have disagreed with at least part of that statement. (As did many of his readers: The article was so popular that there will now be three columns a year devoted to literate programming.)

As Don began explaining the "`WEB`" system, he restated two previously mentioned principles: The correct way to explain a complex thing is to break it into parts and then explain each part; and things should be explained twice (formally and informally). These two principles lead naturally to programs made up of modules that begin with text (informal explanation) and finish with Pascal (formal explanation).

The WEB system allows a programmer to keep one source file that can produce either a typesetting file or a programming language source file, depending on the transforming program used.

Monday's final topic was the "blight on the industry": user manuals. Don would like us to bring in some really stellar examples of bad user manuals. He tried to find some of his favorites but found that they had been improved (or hidden) when he wasn't looking. While he could have brought in the improved manuals, bad examples are much more fun.

He showed a brand-new book, *The AWK Programming Language*, to illustrate a principle often used by the writers of user manuals: Try to write for the absolute novice. He says that many manuals say just that, but then proceed to use jargon that even some experts are uncomfortable with. While the AWK book does not explicitly state this goal, the authors (Aho, Weinberger, Kernighan = AWK) told him that they had this goal in mind.

But the book fails to be comprehensible by novices. It fails because, as Don says, "If you are a person who has been in the field for a long time, you don't realize when you are using jargon." However, Don says that just because the AWK book fails to meet this goal does not mean that it isn't a good book. ("Perhaps the best book in Computer Science published this year.") He explains this by saying, "If you try to write for the novice, you will communicate with the experts—otherwise you communicate with nobody."

§11. Excerpts from class, October 21 [notes by PMR]

Don opened class with the good news that Mary-Claire van Leunen has agreed to help read the term papers and drafts thereof, despite the fact that her name was incorrectly capitalized in last week's notes.

Returning to the subject of "Literate Programming," Don said that it takes a while to find a new style to suit a new system like WEB. When he was trying to write the WEB program in its own language he tore up his first 25 pages of code and started again, having finally found a comfortable style. He digressed to talk about the vicious circle involved in writing a program in its own language. To break it, he hand-simulated the program on itself to produce a Pascal program that could then be used to compile WEB programs. The task was eased because there is obviously no need for error-handling routines when dealing with code that you have to debug anyway. But there is also another kind of bootstrapping going on; you can evolve a style to write these programs only by sitting down and writing programs. Don told us that he wrote WEB in just two months, as it was never intended to be a polished product like TeX.

We spent the rest of the class looking at WEB programs that had been written by undergraduates doing independent research with Don during the Spring. We saw how they had (or had not) adapted to its style. Don said that he had got a lot of feedback and sometimes found it hard to be dispassionate about stylistic questions, but that some things were clearly wrong. He showed us an example that looked for all the world just like a Pascal program; the student had obviously not changed his ways of thinking or writing at all, and so had failed to make any use of the features of the system. The English in his introductory paragraph also left a lot to be desired.

Don showed us his thick book *TEX: The Program*—a listing of the code for TEX, written in WEB. It consists of almost 1400 modules. The guiding principle behind WEB is that each module is introduced at the psychologically right moment. This means that the program can be written in such a way as to motivate the reader, leaving TANGLE to sort everything out later on. [The TANGLE processor converts WEB programs to Pascal programs.] After all, we don't need to worry about motivating the compiler. (Don added the aside that contrary to superstition, the machine doesn't spend most of its time executing those parts of the code that took us the longest to write.) It seems to be true that the best way in which to present program constructs to the reader is to use the same order in which the creator of the program found himself making decisions about them. Don himself always felt it was quite clear what had to be presented next, throughout the entire composition of this huge program. There was at all points a natural order of exposition, and it seems that the natural orderings for reading and writing are very much the same.

The first student hadn't used this new flexibility at all; he had essentially just used WEB to throw in comments here and there.

A general problem of exposition arose: How are we to describe the behavior of a computer program? Do we see the program as essentially autonomous, "running itself," or are we participants in the action? Our attitude to this determines whether we are going to say 'we insert the element in the heap' or 'it inserts the element ... '. Don favours 'we'; at any rate one should be consistent.

Students used descriptors and imperatives for the names of their modules; Don said he favours the latter, as in ⟨ Store the word in the dictionary ⟩, which works much better than ⟨ Stores the word in the dictionary ⟩. On the other hand, where a module is essentially a piece of text with a declarative function—a list of declarations, say—we should use a descriptor to name it: ⟨ Procedures for sorting ⟩.

Incidentally, it is natural to capitalize the first letter of a module name.

One student used the identifier '*FindInNewWords*'. This looks comparatively bad in print: Uppercase letters were not designed to appear immediately following lowercase ones. Since the use of compound nouns is almost inevitable, WEB provides a neat solution. It allows a short underscore to be used to conjoin words like *get_word*. (Since the Pascal compiler will not accept identifiers like this, TANGLE quietly removes the underscore.) Don told us that Jim Dunlap of Digitek, who made some of the best early compilers, invariably used identifiers forty-or-so characters long. The meaning was always quite clear although no comments appeared.

Each module should contain an informal but clear description of what it actually does. A play-by-play account of an algorithm, a simple stepping through of the process, does not qualify. We are trying to convey an intuition of what is going on, so a high-level account is much more helpful.

We saw several modules that were much too long. Don thinks that a dozen lines of code is about the right length for a module. Often he simply recommended that the students cut the offending specimens into several pieces, each of more manageable size. The whole philosophy behind WEB is to break a complex thing into tractable parts, so the code should reflect this. Once you get the idea, you begin writing code this way, and it's easier.

We saw an example in which the student had slipped into "engineerese" in his descriptive text—all conjunctions and no punctuation. This worked for James Joyce, but it doesn't make for good documentation. One student had apparently managed to break WEB—the formatting of **begin**s and **end**s came out all wrong. Heaven knows what he did.

One student put comments after each **end** to show what was being ended, as **end** {**while**}. This is a good idea idea when writing Pascal, but it's unnecessary in WEB. Thus it's a good example of a convention that is no longer appropriate to the new style; when you change style you needn't carry excess baggage along.

Don had more to say about the anthropomorphization of computer systems. Why prompt the user with '`Name of file to process?`' when we can have the computer say '`What file should I process?`'? Don generally likes the use of 'I' by the computer when referring to itself, and thinks this makes it easier for users to conceptualize what is going on. Perhaps humans can think of complex processes best in terms of demons in boxes, so why not acknowledge this? Eliza, the AI program that simulates a certain type of psychiatrist, managed to fool virtually everyone by an extension of this approach. Eliza may or may not be a recommendation for anthropomorphisms, or for psychiatry. There are those, such as Dijkstra, who think such use of 'I' to be a bad thing.

As in the case of maths, don't start a sentence with a symbol. So don't say '*data* assumes that ...'—it can easily be rewritten.

We saw several programs by one student who had developed a very distinctive and (Don thought) colourful style. His prose is littered with phrases like "Oooops! How can we fix this?" and "Now to get down to the nitty-gritty." This stream-of-consciousness style really does seem to motivate reading, and helps infect the reader with the author's obvious enthusiasm. There were a few small nits to pick with this guy though: His descriptions could often be more descriptive. Why not call a variable *caps_range* instead of just *range*? Don also had to point out to him that 'complement' and 'component' are in fact two different words.

In WEB you can declare your variables at any point in the program. Don thinks it is always a good idea to add some comment when you do so, even if only a very cursory explanation is needed.

A note about asterisks: Be warned that typeset asterisks tend to appear higher above the line than typewritten ones, so your multiplication formulæ may come out looking strange. Better to use × for multiplication, and to use a typewriter-style font with body-centered '*' symbols instead of the '*' in normal typographic fonts.

Another freshman was digitizing the Mona Lisa for reasons best known to insiders of Don's research project. Don pointed out that since the program uses a somewhat specialized data structure (the heap) that might be unknown to the readers, the author should keep all the heap routines together in the text so that they can be read as a group while fresh in the reader's mind. In WEB we are not constrained by top-down, bottom-up, or any other order.

This student capitalized the first letter of every word in titles of modules, even 'And' and the like. This looks rather unnatural—it is better to follow the newspaper-headlines convention by leaving such words entirely in lowercase, and even better to capitalize only the first word.

Don thought it a good idea to use typewriter type for hexadecimal numbers, for instance when saying '3F represents 63'. But leave the '63' in normal type. This convention looks appropriate and provides a kind of subliminal type-checking.

The words used in the documentation should match the words used in the formal program—you will only confuse the reader by using two different terms for the same thing.

It's a good idea to develop the habit of putting your **begin**s and **end**s inside the called modules, not putting them in the calling module. That is, do it like this:

>**if** $down = 4$ **then** $\langle$ Punt $\rangle$;
>$\vdots$
>
>$\langle$ Punt $\rangle =$
>   **begin** snap;
>   place;
>   kick;
>   **end**

Not like this:

>**if** $down = 4$ **then**
>   **begin** $\langle$ Punt $\rangle$;
>   **end**
>$\vdots$
>
>$\langle$ Punt $\rangle =$
>   snap;
>   place;
>   kick

Incidentally, appalling bugs will occur if we mix the two conventions!

§12. Excerpts from class, October 23 [notes by PMR]

One of the chief aspects of WEB is to encourage better programming, not just better exposition of programs. For example, many people say that around 25% of any piece of code should be devoted to error handling and user guidance. But this will typically mean that a subroutine might have 15 lines of 'what to do if the data is faulty' followed by one or two lines of 'what to do in the normal course of events'. The subroutine then looks very much like an error-handling routine. This fails to motivate the writer to do a good job; his heart just isn't in the error handling. WEB provides a solution to this. The procedure can have a single line near the beginning that says ⟨ Check if the data is wrong 28 ⟩ and points to another module. Thus the proper focus is maintained: In the main module we have code devoted to handling the normal cases, and elsewhere we have all the error-case instructions. The programmer never feels that he's writing a whole lot of stuff where he'd really much rather be writing something else; in module 28, it feels right to do the best error detection and recovery. Don showed us an example of this from his undergraduate class in which a routine had two references of the form

> **if** ... **then** ... **else** *char_error*

pointing to a very brief error-reporting module.

We looked at a program written by another student who had the temerity to include some comments critical of WEB. Don struck back with the following:

> It is good practice to use italics for the names of variables when they appear in comments.

> Let the variables in the module title correspond to the local parameters in the module itself.

> According to this student's comments, his algorithm uses 'tail recursion'. This is an impressive phrase, helpful in the proper context; but unfortunately that is not the kind of recursion his program uses.

However, Don did grant that his exposition was good and gave a nice intuition of the functions of the modules.

We saw a second program by the same student. It had the usual sprinkling of "wicked whiches"—'which's that should have been 'that's. The purpose of the program was to "enforce" the triangle inequality on a table of data that specified the distances between pairs of large cities in the US. Don commented here that his project (from which these programs came) intends to publish interesting data sets so that researchers in different phases can replicate each other's results. He also observed that a program running on a table of "real data," as here (the actual "official" distances between the cities in question) is a lot more interesting than the same program running on "random data." Returning to the nitty-gritty of the program, Don observed that the student had made a good choice of variable names—for instance '*villains*' for those parts of the data that were causing inconsistencies. This fitted in nicely with the later exposition; he could talk about 'cut throats' and so forth. (Don added that we nearly always find villainy pretty unamusing in real life, but the word makes for a witty exposition in artificial life; the English language has lots of vocabulary just waiting for such applications.)

Don wondered aloud why it is that people talk about "the nth and m^{th} positions" (as this student had) thereby reversing the natural (or at any rate alphabetical) order?

He also pointed to an issue that arises with the move from typewriters to computer typesetting—the fact that we now distinguish between opening and closing quotes. We saw an example where the student had written "main program". To add to the confusion, different languages have different conventions for quotes; in German they appear like this: „The Name of the Rose". How to represent this in a standard ASCII file remains a mystery.

Back to the triangle inequality. Don pointed out that one obvious check for bad data in the distance table follows from the fact that the road distance can not be less than a Great Circle route. ("It could, if you had a tunnel" commented a New Yorker in the audience.) The student had written a nice group of modules based on this fact, and it illustrated the WEB facility of being able to put displayed equations into comments.

"So WEB effectively just does macro substitution?" asked another member of the class. Exactly, said Don. In fact the macros he uses are not very general—they really allow only one parameter. This means he doesn't need a complex parser, but in fact one can do a great deal within this restriction. For instance, it is not difficult to simulate two-parameter macros if we wish.

Someone in the class commented that it seemed a little strange to put variable declarations in a different module from their use. Don said that this was OK as long as they are close to their use, but large procedures should have their local variables "distributed" as the exposition proceeds.

Don recalled that older versions of Algol allowed you to declare a variable in the middle of a block. This fits in nicely with the WEB philosophy, but unfortunately cannot be done in modern Pascal. Indeed, Don became painfully aware of the limitations of Pascal for system programming when he was writing WEB—you can't have an array of file names, for example. He got around them, though, with macros.

One example of improving Pascal via macros is to define (in WEB)

$$string\_type(\#) \equiv \textbf{packed array } [1\mathinner{.\,.}\#] \textbf{ of } char$$

so that you can say things like

$$name\_code\colon string\_type(2)$$

when declaring a two-letter string variable.

At this point, prompted by a note from Tracy, Don announced that 23 copies of the *Handbook for Scholars* had arrived in the Bookstore, with more to come. A resounding cheer echoed throughout Terman.

Don commented that the student had called a certain variable called '*scan*'. Since this variable was essentially a place marker, Don thought that a noun would be much better than a verb—'*place*', perhaps. Let the function determine the part of speech; think of it as a kind of Truth in Naming. Verbs are for procedures, not data.

The last student had written a program to handle graph structures based on encounters between the characters in novels. He too had made the "quote mistake". The student gave a nice characterization of the input and output of the program, using the typewriter font to illustrate data as it appears in a file.

This student also showed a bit of inconsistency in the use of 'it' and 'we' as the personification of his program. We seem to be finding the same old faults over and over now, Don said, so perhaps that indicates that we have found them all. Discuss.

§13. Excerpts from class, October 26 [notes by PMR]

We moved on to the subject of user manuals. Don was disappointed that nobody had responded to his request in a previous lecture to give him glaring examples of bad ones—either they are being much better written these days or we hadn't taken him seriously. So instead, Don produced mini-sized user manuals written by CS graduate students for his class CS 304 earlier this year. The students had had to tackle five weird and wonderful problems in ten weeks; one of the problems had been to design and implement some software and to write a one-page user manual for the 'Digiflash' display system. This is the kind of thing you see in Times Square, and increasingly in bars and post offices, in which news and advertising flows across a sort of dot-matrix screen. In this case, the screen was to be a simple array of 8 by 256 pixels. The students had only two weeks in which to write the system and manual, which were then subject to the ultimate test, the truly Naïve User. The idea was that the user would need no understanding of computers or of writing, but should still be able to use the system to produce a variety of visual effects. The students divided themselves into four teams and so we saw four solutions to the problem.

A common failing was that terminology that seemed perfectly transparent meant nothing to Don's wife: "Menu" and "Scrolling" for example. Such terms are so familiar to CS people, it never occurs to them that these are actually technical terms.

Don went through the solutions in ascending order of competence. The class reaction to this discussion might almost have led one to believe that some of the authors were sitting amongst us.

Don digressed on the subject of 'i.e.'. Is it formal, he asked, or is it part of the language? He confessed that he was considering taking all the 'i.e.'s out of his new book. One thing he does know: You should always put a comma after 'i.e.'. (Except in this instance.)

The first solution could be described as a very *logical* approach, almost an archetypically CS solution. The manual was essentially a hierarchy of definitions. The writers talked about MESSAGES (or MESSAGEs—consistency was not their watchword) when they wanted to say: 'here are objects that have a special meaning for us and whose definition you ought to know'. But, said Don, formal definitions are not the way to explain something to a novice.

This write-up apparently thought the phrase LEFT-INDENTED to be self-explanatory, although it meant 'flush with the left margin'. (Left unindented?) The user was prompted to enter '`Type of message (1-6):`' Why should there be numeric types? Sentences like "And now you should ENTER the data" do nothing to help the user relax—the capitals look too much like DANGER SIGNS.

Don's wife commented that one thing she always needs to know is "How do I get out of a mess if I do something wrong?" Don said that this is something manuals almost never explain—perhaps it never occurs to their authors that somebody will eventually want to stop playing with their program. The solution we were looking at did have a one-line description of how to EXIT, but Don said even this is jargon.

The second solution was Digiflash™. It had a good introductory and motivational paragraph, albeit with a whole crowd of 'which's that should have been 'that's. Unfortunately it claimed that the system was very easy to use and understand—a claim that can rebound by making the user feel stupid. There was a major flaw in the program in that one was expected to hit OPTION-B to enter 'bold' mode, and then OPTION-B again to leave it. Don thought it would be far more natural to type OPTION-N (for 'normal') instead. Option-V was "reverse video"—another jargon word, and why wasn't this OPTION-R? There were some cute options though: 'M' for 'slowly materializing' text, and an assortment of small animal logos that could be made to appear.

The third solution was the DiJKSTra system, so named to keep it sufficiently Dutch (obscure in-joke, please ignore). The authors had a nice use of the phrase 'flashing bar' instead of the more technical 'cursor' (though for some reason they still felt impelled to define the latter as the former), and likewise they said 'hit return' instead of 'enter' (or worse, ENTER). They also kept their sentences nice and short. Another good idea was that the manual invited the user to type '?' to get an online demonstration, thus sparing us a painful description of such arcane concepts as boldface italic reverse video fade-in mode and incidentally helping to keep the manual concise. If a picture is truly worth a thousand words, said somebody, then an animated demonstration must be worth at least a paragraph. One problem with this system was that the user is prompted for five or so parameters every time he enters a new line, and the defaults are fixed. Wouldn't it be better, asked Don, to default to the style used for the preceding line?

The last solution (though not even typeset, much less TEXed) Don declared to be the best. There was a good overview and a step-by-step description of the system; very friendly looking. Crisp sentences. Easy to skim. Helpful redundancy and diagrams. Don said that there's really nothing much you can do about the reader who insists on starting at a random point in the middle of a text. When he surreptitiously watches people looking at his books in the bookstore, he notices that they always start in the middle somewhere, not at the preface where he wanted them to read first.*

There was a good use of a symbol in the text to indicate the control key, and likewise diagrams of the keyboard to explain which keys to use for left, right, next message, previous message, etc. It was also good to emphasize that the control key must actually be held

---

* "As for those readers who do not know how to study my composition, no author can accompany his book wherever it goes and allow only certain persons to study it." — Maimonides

down while another key is typed (that is, they are not simply typed successively). Perhaps the main flaw was that the user was expected to realize that 'up' meant, in effect, 'go to the previous message'; and 'down', 'go to the next message'. To those unfamiliar with full-screen editors, this mightn't be obvious. There was a nice use of icons to describe scrolling up, down, left, or right though. One obscurity was the advice

> `THIS IS VISIBLE HERE`   `BUT NOT HERE`

Don declared that he didn't know what this was supposed to mean; it would be a lot better to say 'Extra long messages can be seen if you make them move'.

It's good to have plenty of comments like 'Good luck!' and 'Enjoy!' scattered here and there. But Don thought the phrase 'this system has been carefully redesigned not to bite' hardly reassuring.

§14. Excerpts from class, October 26 (continued) [notes by PMR]

In Don's mailbox today he found galley proofs from the ACM, to be corrected and returned within 48 hours of this time two weeks ago. Unfazed by this injunction he went over the text with us. The Algol programs seemed to be laid out properly. There were occasional cryptic marginal notes: 'Bad proof, Camera copy OK'. He took this to mean that his copy was made by a laser printer instead of a phototypesetting machine. We learned that 'Au' means not gold but 'author' in the copy-editing world. The copy editor had substituted 'cleverer' for Don's 'more clever', citing Fowler. Don sighed and recalled the occasion that *Scientific American* had replaced his 'more common' with 'commoner'. It was noticeable that the copy editor was not going to change anything without Don's specific approval—not even removing the first 'of' in '...several possible of values of the variable n ...'. Don told us that at the moment all papers are re-typed by the publishers, except for one or two AI journals that have used TEX for several years. But next year a math journal will be adopting a policy in which the author's text is manipulated electronically throughout the whole process. This should speed publication and reduce errors and costs.

Some of the notes in the galley were signed 'Ptr', that is 'printer', and asked 'OK?'. Don answers affirmatively by circling the 'OK'. At one point he was asked to sanction the insertion of a whole new sentence. Apparently he had made reference to Figure 14 before Figure 13, and his approval was sought to make an extra comment first about 'Figures 13–16'. (The extra comment was wrong but fixable.)

The publishers also insisted on more details in his bibliography. They wanted to know, for example, exactly where and when a conference had taken place. Someone in the class pointed out that Mary-Claire van Leunen recommends omitting the location of conferences. Don replied that libraries often nowadays index conferences by city for those poor souls who can remember nothing else about them; so such information was useful. He observed that people have a great tendency to copy citation information blindly into their own papers, and so errors propagate unchecked. When Elwyn Berlekamp wrote his book on coding theory, he found that nearly half the information in bibliographies of papers was wrong! Don wrapped up the galley proof discussion by showing us a few tables of (somewhat) standardized proof-readers' symbols.

§15. Excerpts from class, October 30 [notes by PMR]

Today Don spoke about the refereeing process. A paper submitted to an academic journal is usually passed to one or more referees by the editor of the journal. Each referee is intended to be an expert in the relevant field, and thus in a position to tell the editor whether or not the paper merits publication. Don pointed out that many of us will one day find our papers being subject to just this scrutiny; and some of us will certainly be asked to assess other people's papers ourselves.

Don talked about his now famous research on "The Toilet Paper Problem." This was first published in the MONTHLY, and as Don pointed out to the Editor in his cover letter, many of its readers probably keep their copies in the bathroom anyway. The editor (Halmos) replied a little gravely that "jokes are dangerous in our journal", and asked Don to think twice about the scatological references. Don did agree to change the section names—which originally continued the pun with such headings as 'An absorbing barrier', 'A process of elimination', and 'Residues'—to innocuous equivalents, but kept the title intact. In justification of this, Don pointed out to the editor that two talks had already been given on his results under this title, and that the material had been widely circulated and discussed. "Your toilet paper is accepted" replied Halmos. Don confessed that he still has occasional doubts when he catches sight of the title amongst his papers, but the deed is done now. Still, it did get reasonably good reviews, even in Russia.

Don showed us an article entitled 'Rules for Referees' by Forscher, published in *Science* (October 15, 1965). These rules constitute a rather traditional view, Don said, and emphasize the legal rights and responsibilities of all concerned. Don thought that this seems a lot more oriented to the advancement of careers rather than of science as such; the right reason to publish is to build upon the results of others and provide a foundation for future research. It is a sad truth, said Don, that an editor can all too easily find himself spending a great deal of time dealing with those authors whose papers don't merit publication, for it is usually very hard to convince them of the fact. Rebuttals are followed by counter-rebuttals, and so on. But fortunately this doesn't happen so often that the whole business of science gets bogged down.

The referee is conventionally regarded as a sort of "expert witness," whose task is to tell the editor whether the paper deserves to be published or not. The first criterion should be originality; is the material presented a genuine advance on previous work?

Don urged referees to see their primary responsibility as being to authors and readers, not just to editors. Don himself decided long ago that he would put more of his efforts into refereeing papers before their publication than into reviewing published papers. Don hoped that he could thus do his bit to encourage high standards of writing in Computer Science and help the field win respect. These days there are more good people to go around, both in refereeing and reviewing.

In the 1960's Don circulated a list of 'Hints to referees' to try to encourage good practice. He would like to show us that list, but not a copy can be found. Don has written to some of the people to whom he sent it, so it is possible that a copy will turn up before the end of the quarter.

Don disagreed with our guest speaker, Herb Wilf, who had said that he would tolerate more stylistic lapses in the *Journal of Algorithms* than in the MONTHLY. Authors, thought Don, should always be encouraged to do better; he could recall only a single occasion when, as a referee or editor, he could recommend no improvements at all. (The author in this case was George Collins writing for the ACM journal.) Let us publish journals to be proud of, he said. This was sadly not true of Computer Science in the early 60s. Some published results were just plain wrong; or a correct result was incorrectly proved; or a paper simply contained no results at all! Contrast this state of affairs, said Don, to the math journals that were published in the 20s and 30s—leafing through them at random we see a host of familiar names and theorems, because so much of what was written then was polished, significant, and worth reprinting in textbooks. The same could not be said of today's efforts—perhaps we have grown increasingly tolerant of substandard work.

Referees should try to be teachers, said Don. The author you criticize today will be writing another paper tomorrow, so try to help him improve his writing. Unfortunately, referees will often be over-critical and make quite tasteless comments on papers, knowing that they do so under a cloak of anonymity. This only angers the author and he learns nothing. Try to supply constructive criticism, Don urged. These human issues are not discussed in Forscher's 'Rules'.

In addition, the referee can contribute to the technical quality of a paper by giving references to related work of which the author was apparently unaware, or improving the results. Don himself has contributed results anonymously to papers—more than one author has had to add a footnote: "My thanks to the referee for Theorems 4, 5, and 6." Don was always pleased to feel that by doing this the image of the journal was improved. A journal should be seen as a source of wisdom, so let us be cooperative toward this end, not legalistic.

How should one choose a journal to which to submit a paper? Don thought the answer is to look for the one with the best referees, not the one with the least critical editor. After all, an author presumably wants to know whether he has really made a contribution to his field. So find a journal that has handled papers on related subjects.

Someone asked whether the letters that appear in journals are also refereed. Don said that sometimes they are, sometimes not. There is often nothing to distinguish letters from short papers.

Some journals do not use referees at all. Their readership must be willing to wade through a great deal of nonsense. The ACM did at one time have plans to publish an unrefereed journal, but to Don's relief those plans never came to fruition.

At this point Don confessed to a sneaky trick he had pulled way back in the 60s. At that time he had just begun to edit the programming languages material for the *Communications of the ACM* and the *Journal of the ACM*. He had no way of knowing which of his referees were any good, so in an effort to calibrate them he sent all a copy of the same paper and solicited their opinions. Don had already refereed the paper himself, of course, and found it a very badly written exposition of a very interesting algorithm (due to someone besides the author). As such, it was certainly worthy of the referees' study.

We looked at some of the results. One commentator simply went through line by line, listing his complaints point by point. Another made much more general comments: "A paper with this title should contain (1) a complete algorithm; (2) a proof or at least a convincing explanation of correctness; (3) a statement of limitations on the algorithm's applicability. None of these can be found here." A third said that the paper contained little that was new, and supplied a substantial bibliography for the author to go away and study. The next referee liked the algorithm and recommended the paper for publication. Don was surprised; he had mistakenly thought that this referee had originally invented the algorithm himself! Another critic dismissed the paper as 'incredibly poorly written'. Another commented it was not a paper to be read, but rather a puzzle to be solved.

Don told us that as a result of his experiment, the algorithm actually became quite well-known.

On one occasion Don ripped into a paper with a long report on its failings, and was later told by the author that those constructive comments had changed his life: The author had resolved that from then on he was going to study writing and give a lot of attention to exposition. This nameless individual went on to become a renowned professor at a great (but here equally nameless) university, and an editor of a fine journal.

In answer to a question, Don said that if the content of a paper was obviously bad, he would not spend time reviewing the grammar. But in studying a paper that really has something to say, then he would also try to ensure that it was said as well as possible.

Don showed us some referees' reports on one of his recent papers. The editor had told him that these were 'mostly positive'—in fact two were in favour and one against. The referees in this case had been asked to answer a specific list of questions about his paper—Don said that this tedious format might at least cause a referee to consider issues he might otherwise have forgotten about. The referees did agree that Don hadn't made enough reference to earlier work in the subject. This didn't surprise him; the paper was his first venture into an unfamiliar field. The referees were helpful enough to comment now and then that they had particularly enjoyed certain sections, and they provided a whole slew of references to other work—references that Don said had led to some new ideas. They were also able to point out subtle technical errors; Don had to write a program to convince himself that one in particular of these criticisms was valid. Finally, we were amused to see that the referees were asked to assign an overall rating to the paper by checking one of a series of boxes, ranging from (as the most lavish praise) "accept", down through "accept with major modifications" and "accept with minor changes" to "Reject: submit to _____". When checking this last box (and most damning indictment), the referee was asked to suggest a less prestigious journal that might publish such inferior work. By such a downward filtering, even the most appalling paper stands some chance of finding its place in the pages of what Dijkstra has characterised a "Write-Only Journal." With four new scientific papers being published every minute throughout the world, we can rest assured that many do so.

§16. Excerpts from class, November 2 [notes by TLL]

Today's handout, "Hints for Referees" by D. Knuth (see §17 below), could have been subtitled "Ask and ye shall receive." Last Friday Don mentioned in class that he could find no copy of this document, but when he returned to his office immediately after class he found it sitting on his desk. (To be truthful, he thinks this copy has gone through a few revisions since it left his hands; he no longer recognizes the style of all the comments.)

Before demonstrating to us how highly he esteems referees and the lengths to which he will go in order to get referees, Don told us to note an important date on our calendar: On Wednesday, November 18, we are to turn in the first drafts of our Term Papers ("The closer to the final version, the better").

The identities of the referees for a journal paper are usually hidden from the author. Is the identity of the author ever hidden from the referees? In some few journals, yes. Don is well aware that the name written just below the title of a paper can strongly effect the reader's reaction, so he submitted a journal paper using the sobriquet, Ursula N. Owens. (Those of us who have read Agatha Christie's *And Then There Were None* realize that his near-use of the name U. N. Owen is a pleasant allusion.)

Don doesn't always resort to pseudonyms, but neither does he always get his papers refereed. On occasion he has recruited his own referees when he found out that his target publication was supplying none. As an example, his paper on Literate Programming for the British *Computer Journal* generated no referee reports (and no feedback of any kind); they went right into print.

Don repeatedly stated how invaluable he found "feedback from a motivated reader." He showed us the comments that "Ursula" got on her paper, and they were pertinent in more than one way. The referee found typographic errors and suggested notation changes, as well as finding errors where there were none present. The last set of comments were more important than they might at first seem because they pointed out where Don's presentation was misleading or overly subtle.

In another example, the referee significantly improved one of the theorems while remaining anonymous. Instead of being content with an acknowledgement to an anonymous contributor, such a referee could be jealous and publish his own competing paper.

In contrast to such substantive contributions, Don showed us another example with suggestions that he called "facile generalizations" (terminology attributed to Pólya): generalizations that are merely mechanical manipulations of a given argument without creating new insight.

Don says that refereeing is a "cooperative effort—a correspondence between tens of thousands of world authorities," and he is perfectly willing to exploit the system by letting referees improve his papers as he helps with theirs.

He showed us a series of letters passing between himself and the *Journal of Number Theory*. He had produced a result that seemed novel (could not be found by exploring the standard pathways in the Math Library), but since Number Theory is not his field of expertise, Don was unwilling to claim that the result was not a duplication. He told this to the editors of the journal and asked for feedback. ("I put in a lot of time reviewing other people's papers. This is my chance to get some of that time back.")

The referee reports on that paper found references that Don "couldn't have found in a million years." The results were similar but not identical, so the referee offered to check with a famous Russian expert. As Don was availing himself of this offer, someone else was publishing on the same subject. ("You have to decide, do you want speedy publishing or rigorous checking?")

Finally, he showed us two examples that dealt with ambiguity. In the first, he and David Fuchs had written a paper entitled "Optimal Font Caching." One of the refereees pointed out that this paper could be about the caching of optimal fonts, or the best of all possible caching mechanisms for fonts. An analogous title "Common Sense Amplifiers" was cited. (Don and Dave solved this problem by changing the title to "Optimal Prepaging and Font Caching.") In the second, he had to cope with the *IEEE Journal on Coding Theory*'s penchant for writing out the words 'one' and 'zero' for the symbols '1' and '0'. Since 'one' is an English pronoun, he was forced to use the word 'unity' in one place to make the text unambiguous.

§17.   Hints for referees
                     (please keep in your file)
                                                              D. Knuth

In a relatively new field such as computing there is bound to be a lot of trash published since there are too few people available to recognize the poor quality of much of the material. But this discourages people in the computing profession from reading the literature and causes a poor image for the profession in the eyes of others. The only way to prevent this is to have a strong refereeing system. Although the job of refereeing is not simple, it is an important responsibility, nearly the most important thing anyone could be doing for the field of computer science.

Papers generally will fall into the following categories:

1. Publish essentially as is; the only changes necessary are very simple typographical matters which can be changed by the editor.

2. Publish after author's minor revision; the referee suggests points which must be changed before the paper meets the standards for publication.

3. Publish only if the author makes major revisions. (Perhaps the paper is much too long or is badly written. The revised paper will be refereed again.)

4. Reject. (There is nothing salvageable.)

The goals of a referee are to keep the quality of publication as high as possible and also to help the author to produce better papers in the future. Your referee's report should be designed to give the author the maximum benefit, yet not compromise on quality. Try to get every author to put out the best paper he is capable of writing; a paper rarely falls in category 1 above. Never put a paper in category 1, if you feel the author can do better, even if the paper as it stands is reasonably good! A paper should only be put into category 3, if the substance of the paper is considered significant enough to warrant the additional amount of labor to rewrite and reconsider the paper.

To judge the publishability of the paper you certainly know what is good and what is bad but the following brief list is included here anyway.

(a) The paper should contribute to the state of the art and/or should be a good expository paper. If it is purely expository it should be clearly designated as such.

(b) All technical material must be accurate (e.g., no incorrect equations, etc.). A referee should check this carefully.

(c) The article must be understandable, readable, and written in good English style.

(d) The bibliography should be adequate.

It is tempting to postpone refereeing tasks by putting the paper aside for a few days. But it takes no longer to do it today than it will in a week's time. If you feel that you are for some reason unable to referee the paper please return it immediately. Otherwise, the referee's report is expected in no more than four weeks. Remember that the refereeing cycle is "critical path time" in the publication process.

Return the manuscript to the editor; please don't mark it up. You should submit the report in duplicate. Remember that one copy will be sent directly to the author; it is up to you whether you want to mention your name on it or not. If you desire, you may write an accompanying letter to the Editor which, of course, will not be passed on the author. This letter, however, must not constitute the referee's report.

§18. Excerpts from class, November 4 [notes by PMR]

Today, Don said, we are going to talk about the use of pictures and illustrations in mathematical writing, and about the problem of "getting across the feeling of complicated algorithms."

But first, by popular demand, Don showed us his first publication. This was a description of a system of weights and measures known as the Potrzebie System, which appeared in the pages of *MAD* magazine in 1957. Any resemblance to the Metric System is purely coincidental. It an extremely natural and logical system, Don told us. For example, the units of time were named after the editors of *MAD* (the new editors substituted their own names). He felt there was also a need for new units of counting, and so coined the MAD; 48 things constitute one MAD (or 49, a baker's MAD). Don didn't publish a better illustrated work until *The TEXbook*, he claimed, nor another paid one until he wrote for ACM *Computing Surveys* some 12 years later. *MAD* forked over no less than $25 for this research paper, no mean sum thirty years ago. 'The Potrzebie System' still heads the list of publications on his C.V.

MAD inexplicably declined Don's second article, "RUNCIBLE: Algorithmic Translation on a Limited Computer," which was picked up by *Communications of the ACM* in 1959. Perhaps this was because it contained what even Don admits is probably one of the worst "spaghetti" flow charts ever drawn. Not only does the chart attempt to illustrate the entire algorithm, but it contains an error (a misdirected arrow). The article included a play-by-play account of the algorithm, which helped ameliorate the obscurity of the chart. Back in those days, Don now admits, he didn't know any better. Likewise, full of youthful enthusiasm at being able to communicate improvements on a previously published algorithm (Don was a Junior then), he failed to mention his co-authors in the paper; Don did the writing but other students contributed illustrations and most of the ideas of the algorithm. At the time he had no notion there was academic prestige to be gained through publication, Don confessed. This is, he said, a common mistake among young authors who frequently overlook proper acknowledgements in their haste to get the news out. At the other extreme, he recalled, Paul Erdős once cited a railroad car porter as a co-author.

Diagrams are good if they are kept small, said Don. As an example of a useful one that is *not* small, he showed us a fold-out syntax chart for a slightly extended version of the Algol 60 language. It does convey quite a good impression of what the language is, and gives computer scientists something to hang on the wall where chemists put their Periodic Tables.

Don's "Programming Pearls" article came up again. He had ended that paper with the observation that the only fair test of his WEB system would be this: Someone should provide a challenge problem, and Don would use WEB to write an ultra-elegant solution to it. Jon Bentley responded to this challenge; he devised such a problem and invited Don to submit his solution for publication. Holding Don to his claim that WEB programs should be works of literature, Jon then published the solution along with a literary critique. In this review Jon commented that Don could have eased the exposition of his data structure with a suitable diagram. Don agreed that this would have helped the reader get a handle on it (he had described the data structure in words only). He told us that diagrams were actually quite easy to do in WEB, a claim that was greeted with a certain skeptical laughter from the class (all doubtless recalling hours spent wasted trying to get tables *just* right).

Referring again to his 'optimal prepaging' paper (which included a diagram in which two approximately diagonal lines crawled across the page, touching occasionally to indicate a page fault) Don told us that the referee had complained that the figure was too detailed. Don disagreed with this, saying that the detail was there for those who want to see it, but could easily be ignored by those who don't. Don confessed that he always has been very concerned with the minutiæ of his subject, and seldom thought any detail too trifling to bother with.

Don discussed a paper he had written with Michael Plass on TEX's algorithm for placing line-breaks in a paragraph [*Software—Practice & Experience* **11** (1981), 1119–1184]. The main difficulty writing the paper was: How to describe the problem and the new algorithm? First of all, they chose a paragraph from one of Grimm's Fairy Tales as "test data" with which to illustrate the process. As Don remarked once before, it is better to use "real" data than "sample data" that have in fact been cooked up solely to use as an example. [*Grimm's Fairy Tales*, along with the text of *Harold and Maude*, are kept on-line on SAIL, an ancient and eccentric CSD computer.] Corresponding to each line of any right-and-left-justified paragraph is a real number, positive or negative, indicating the degree to which the line had to be stretched or compressed to fit the space exactly. In his paper, Don prints these numbers in a column beside his typeset paragraph. Don used a couple of lines of the paper itself to show how bad it looks if these adjustments are too extreme (and of course had to tell the printers that this was a deliberate mistake, lest they "correct" it).

Don outlined three basic algorithms: first fit (which essentially packs the text as tightly as possible one line at a time); best fit (which can loosen it up if this works better, but still works line by line); and optimum fit (optimal in the sense that it minimizes the sum of the "demerits" earned by the various distortions of each line, taken over the paragraph as a whole). To describe this last algorithm, Don drew a diagram. It is essentially a graph, each node on level p corresponding to a different word after which the pth line might be broken. Edges run between nodes on successive levels, and are labelled by the demerits scored by the line of text they define. The problem of finding an optimal fit thus reduces to finding a least-cost path from the top to the bottom node; well-understood search techniques can be used for this. Don commented that certain "demerit-cutoffs" will limit the number of nodes on each level and thus speed the algorithm. This means that a solution in which one very distorted line permits all the rest to be displayed perfectly might be missed.

If the above account is opaque, it only goes to show why diagrams can be so useful.

The article includes histograms to illustrate how frequently TEX generates more-or-less distorted lines of text. As he explained, this was biased by the fact that he would usually re-write any particularly ugly paragraph. A second histogram confirmed that the text was considerably more distorted when it hadn't been hand-crafted to the line-width that TEX was generating, yet the new algorithm was significantly better than Brands X and Y.

Finally, we saw an old Bible whose printers were so keen to fill out the page width that they inserted strings of o's to fill up any gaps.

Don found many illustrative illustrations in the book *The Visual Display of Quantatative Information* by Tufte. He also recommended *How to Lie with Statistics* by Huff, which advises (for example) that if you would impress your populace with the dazzling success of the Five-Year Plan in increasing wheat production by 17%, then draw two sacks, the first 6 cm and the second 7 cm tall. The perceived increase, of course, corresponds to the apparent volumes of the sacks, and 7^3 is 58% larger than 6^3. ...

Don referred to Terry Winograd's book *Language as a Cognitive Process*. Algorithms for parsing English sentences are there illustrated as charts defining augmented transition networks or ATN's—nodes correspond to internal states, edges are transitions between states and correspond to individual words. Winograd also has a nice use of nested diagrams—boxes within boxes—to replace the more traditional tree diagrams.

We saw a scattergram of smiley-faces of somewhat indeterminate significance; a wit in SITN projected Don's amongst them. The idea is that several dimensions of numeric data can be used to control features on these faces; humans are supposedly wired to read nuances in facial expressions quite easily.

Don showed us a table from his *Art of Computer Programming* that listed the many, many states of the Caltech elevator. He said he wished that he'd been able to dream up a diagram to capture that example more neatly: A listing of events is the best way he knows to convey the essential features of asynchonous processes.

The third Volume of this tome does contain a large fold-out illustration comparing the performances of various sort-on-tapes algorithms. Certain subtleties arise from overlaps, rewinds, and buffering that tend to elude conventional algorithmic analysis. Don's diagram neatly captures these, and clearly shows that certain sophisticated algorithms—one was even patented by its author—are in fact slower than traditional methods. Unanticipated rewind times can cause significant slow-downs, and the chart shows why.

§19. Excerpts from class, November 6 [notes by TLL]

We spent the first half of class examining the solutions to a homework assignment (see §20 below). Don says that the solutions were surprisingly good (see §21).

One of the proofs described in that section contains illustrations in four colors. Don says that color can be used effectively in talks, but usually not in papers (for that matter, Leslie Lamport says that proofs should never be presented in talks, but only in papers). Technical illustrations, even without four colors, cause no end of trouble: Don says that the amount of work involved in preparing a paper for publication is proportional to the cube of the number of illustrations. But they are indispensable in many cases.

Don showed us several of the illustrations, charts, and tables from *The Art of Computer Programming*, Volume 3, and recounted the difficulties in choosing clear methods of presenting his ideas. He also mentioned some technical and artistic problems that he had with an illustration: At what angle should the truncated octahedron on page 13 be displayed?

His books contain some numerical tables ("which are sometimes thought to be unenlightening"); Don says that they can sometimes present ideas that can't be demonstrated graphically (such as numbers oscillating about 2 with period 2π, page 41). Diagrams with accompanying text are also used. Don made sure that the final text was arranged opposite the diagrams to which it refers.

The book contains a running example of how 16 particular numbers are sorted by dozens of different algorithms. Each algorithm leads to a different graphical presentation of the sorting activities on those numbers (pages 77, 82, 84, 97, 98, 106, 110, 113, 115, 124, 140, 143, 147, 151, 161, 165, 166, 172, 175, 205, 251, 253, 254, 359).

§20. A Homework Problem

The Appendix to Gillman's book takes a paper that has horrible notation and simplifies it greatly. Your assignment is to take Gillman's simplification and produce something simpler yet. Aim for notation that needs no double subscripts or subscripted superscripts. This assignment will be graded! Please take time to do your best.

Here is a statement of Gillman's simplification. This is your starting point. What is the best way to present Sierpiński's theorem?

Lemma. *There is a one-to-one correspondence between the set of all real numbers α and the set of all pairs $(\langle n_k \rangle, \langle t_k \rangle)$, where $\langle n_k \rangle_{k \geq 1}$ is an increasing sequence of positive integers and $\langle t_k \rangle_{k \geq 1}$ is a sequence of real numbers.*

Notation. The sequences $\langle n_k \rangle$ and $\langle t_k \rangle$ corresponding to α are called $\langle n_k^\alpha \rangle$ and $\langle t_k^\alpha \rangle$. The set of real numbers is called **R**.

Theorem. *Assume that $\langle A_\alpha \rangle_{\alpha \in \mathbf{R}}$ is a family of countably infinite subsets of $\mathbf{R}$ such that, for $\alpha \neq \beta$, either $\alpha \in A_\beta$ or $\beta \in A_\alpha$. Then there is a sequence of functions $f_n : \mathbf{R} \to \mathbf{R}$ such that, if S is any uncountable subset of $\mathbf{R}$, we have $f_n(S) = \mathbf{R}$ for all but finitely many f_n.*

Proof. Let the countable set A_α consist of the real numbers

$$\{\alpha_1, \alpha_2, \alpha_3, \ldots\}.$$

If α is any real number, define an increasing sequence of positive integers $\langle l_k^\alpha \rangle$ by letting $l_1^\alpha = n_1^{\alpha_1}$ and then, after l_{k-1}^α has been defined, letting l_k^α be the least integer in the sequence $\langle n_1^{\alpha_k}, n_2^{\alpha_k}, \ldots \rangle$ that is greater than l_{k-1}^α.

Let f_n be the function on real numbers defined by the rule

$$f_n(\alpha) = \begin{cases} t_n^{\alpha_k}, & \text{if } n = l_k^\alpha \text{ for some } k \geq 1; \\ \alpha, & \text{otherwise.} \end{cases}$$

We will show that the sequence of functions f_n satisfies the theorem, by proving that any set S for which infinitely many n have $f_n(S) \neq \mathbf{R}$ must be countable.

Suppose, therefore, that $\langle n_k \rangle$ is an increasing sequence of integers and that $\langle t_k \rangle$ is a sequence of real numbers such that

$$t_{n_k} \notin f_{n_k}(S), \qquad \text{for all } k \geq 1.$$

Let $t_j = 0$ if j is not one of the numbers $\{n_1, n_2, \ldots\}$. By the lemma, there's a real number β such that $n_k = n_k^\beta$ and $t_k = t_k^\beta$ for all k.

Let α be any real number $\neq \beta$ such that $\alpha \notin A_\beta$. We will prove that $\alpha \notin S$; this will prove the theorem, because all elements of S must then lie in the countable set $A_\beta \cup \{\beta\}$.

By hypothesis, $\beta \in A_\alpha$. Hence we have $\beta = \alpha_k$ for some k. If we set $n = l_k^\alpha$, we know by the definition of f_n that

$$f_n(\alpha) = t_n^{\alpha_k} = t_n^\beta = t_n.$$

But the construction of l_k^α tells us that $n = n_j^{\alpha_k} = n_j^\beta = n_j$ for some j. Therefore

$$f_{n_j}(\alpha) = t_{n_j}.$$

We chose $t_{n_j} \notin f_{n_j}(S)$, hence $\alpha \notin S$. ∎

[Here are additional excerpts from TLL's classnotes for October 16, when the homework problem was handed out:] The first thing that we learned in class today was that now would be a good time to buy Leonard Gillman's book (*Writing Mathematics Well*). Not only have several copies (finally) arrived at the bookstore, but Don has given us a homework assignment straight out of the Appendix of this book.

The assignment (which is due on Friday, October 30th) is to take the "simplified version" of the proof in Gillman's case study and to simplify it still further. The main simplifying principle is to minimize subscripts and superscripts. When we are done, there should be no subscripted subscripts and no subscripted superscripts. As Don said, "Try to recast the proof so that the idea of the proof remains the same, but the proof gets shorter."

The original proof was written by Sierpiński. Don told us that Sierpiński was a great mathematician who wrote several papers cited in *Concrete Mathematics*, from the year 1909 as well as 1959. But the notation in Sierpiński's original proof quoted by Gillman was so complicated that it confused even him: His proof contained an error that was found by another mathematician (after publication).

While the mathematics used in the proof is not trivial, it uses only functions and sets and should be accessible to us. (This is not to say that it is immediately obvious.) Anyone who is uncomfortable with what sets are, what it means for a set to be countable, or what a one-to-one correspondence is, may need some help with this assignment. Don recommended visiting the TAs during office hours as a good first step for those who feel they need help. (It might also help to remember that Don says, "It's not necessary to understand the proof completely in order to do this assignment.")

Don't worry if the hypothesis of the theorem seems pretty wild; it is pretty wild. It implies the "Continuum Hypothesis." The Continuum Hypothesis states that there are no infinities between the countably infinite (the cardinality of the integers) and the continuum (the cardinality of the real numbers). From 1900–1960, the truth or falsity of the Continuum Hypothesis was one of the most famous unsolved problems of mathematics; Sierpiński published his paper as a step toward solving that problem. Kurt Gödel proved in 1938 that the Continuum Hypothesis is consistent with standard set theory; Paul Cohen of Stanford proved 25 years later that the negation of the Continuum Hypothesis is also consistent. Thus we know now that the hypothesis can be neither proved nor disproved.

[Here are additional excerpts from PMR's classnotes for October 23:] The homework assignment is due a week from today, Don said; so do it as well as possible, and let's not have any excuses!

November 9, 1987

Dear Prof. Gillman,

I'm teaching a class called 'Mathematical Writing' this quarter, and I learned about your excellent MAA booklet just in time to adopt it as a required text for this course. We have gotten much from it, and we thank you for having the inspiration and taking the time to write it.

You may be interested in the enclosed homework problem that I gave, based on the Sierpiński paper discussed in your Appendix. The students' solutions were very instructive for me.

§21. Solutions to the Homework Assignment

Most students pleased the instructor by handling this assignment rather well. Either you already knew a lot about writing, or you have learned something this quarter; in any case the exercise seems to have been good practice.

Several answers or excerpts from answers are attached. First is Solution A, an unexpurgated draft that was written by your instructor before handing out the assignment. The main idea here is to "hold back" before enumerating the elements of a set; you can say that S is countable without writing $S = \{s_1, s_2, \dots\}$. This solution also simplifies Sierpiński's proof in minor ways. For example, it's not necessary to have the hypothesis $\alpha \ne \beta$ to conclude that $\alpha \in A_\beta$ or $\beta \in A_\alpha$, because the existence of a family A_α that satisfies Sierpiński's more complicated hypothesis is equivalent to the existence of a family that satisfies the simplified one.

The grader objected to the last sentence in the first paragraph of my proof. He asks, "Has some 'initialization' of L_α been omitted?" He apparently wants $k = 1$ to be singled out as a special case, for more effective exposition. The sentence makes perfectly good sense to me, but maybe there should be a concession to readers who are unaccustomed to empty constraints.

Solution B introduces two nice techniques of a different kind. First, the lemma becomes a sequence of ordered pairs instead of an ordered pair of sequences. Second, the need for a notational correspondence between α and the corresponding sequence is avoided by just using English words, saying that one is the *counterpart* of the other. In other words, we can hold back in giving notations for a correspondence, since plain words are sufficient (even better at times).

Solution B also "factors" the proof into two parts, one that describes a subgoal (the crucial property that the functions f_n will possess) and one that applies the coup de grace. Much less must be kept in mind when you read a factored proof, because the two pieces have a simple interface. Moreover, the reader is told that the proof is "essentially a diagonalization technique"; this statement gives an extremely helpful orientation. It is no wonder that the grader found Solution B easier to understand than Solution A.

Solution C is by another student who found words superior to notation in this case.

Solution D cannot be shown in full because it contains seven illustrations, some of which are in four colors. But the excerpts that are shown do capture its expository flavor.

A combination of the ideas from all these solutions would lead to a truly perspicacious proof of Sierpiński's theorem.

Solution A

Lemma. *There is a one-to-one correspondence between the set of all real numbers α and the set of all pairs (N, T), where N is a countable set of integers and T is a sequence of real numbers.*

Notation. The set N corresponding to α is called N_α, and the sequence T is called $(\alpha_1, \alpha_2, \ldots)$. The set of real numbers is called $\mathbf{R}$.

Theorem. *Assume that there is an uncountable family of countable subsets A_α, one for each real number α, with the property that either $\alpha \in A_\beta$ or $\beta \in A_\alpha$ for all real α and β. Then there exists a countable family F of functions $f : \mathbf{R} \to \mathbf{R}$ such that, if S is any uncountable subset of $\mathbf{R}$, we have $f(S) = \mathbf{R}$ for all but finitely many $f \in F$.*

Proof. If α is any real number, we can construct a countable set of integers L_α as follows: For $k = 1, 2, \ldots$, let β be the kth element of A_α, in some enumeration of this countable set. Include in L_α any element of N_β that's not already present in L_α because of the first $k - 1$ elements of A_α.

Now let $F = \{f_1, f_2, \ldots\}$ be the countable set of functions defined for all real α as follows:

$$f_n(\alpha) = \begin{cases} \beta_n, & \text{if } n \in L_\alpha \text{ and } n \text{ corresponds to } \beta \in A_\alpha; \\ \alpha, & \text{if } n \notin L_\alpha. \end{cases}$$

We will show that F satisfies the theorem, by proving that any given set $S \subseteq R$ is countable whenever $\{\, n \mid f_n(S) \neq \mathbf{R} \,\}$ is infinite.

Let S be a set such that $N = \{\, n \mid f_n(S) \neq \mathbf{R} \,\}$ is infinite, and suppose that

$$t_n \notin f_n(S), \qquad \text{for all } n \in N.$$

Let $t_n = 0$ for $n \notin N$. By the lemma, there is a real number β such that $N = N_\beta$ and $(t_1, t_2, \ldots) = (\beta_1, \beta_2, \ldots)$.

Let α be any real number such that $\alpha \notin A_\beta$. We will prove that $\alpha \notin S$; this will prove the theorem, because all elements of S then must lie in the countable set A_β.

By hypothesis, $\beta \in A_\alpha$. Hence there is some $n \in L_\alpha$ corresponding to β, and $f_n(\alpha) = \beta_n$ by definition of f_n. Also $n \in N_\beta = N$, by the construction of L_α. But $\beta_n = t_n \notin f_n(S)$, so α cannot be in S. ∎

Solution B
Sierpiński's Theorem

Lemma. *There is a one-to-one correspondence between the set of all real numbers α and the set of all sequences of ordered pairs $\langle(n_l, t_l)\rangle_{l \geq 1}$, where the first components $\langle n_l \rangle$ form an increasing sequence of positive integers and the second components $\langle t_l \rangle$ form a sequence of real numbers.*

We shall call the sequence of ordered pairs corresponding to α the *counterpart* of α, and vice versa.

Theorem. *Suppose that there exists a family of countably infinite subsets of the reals $\mathbf{R}$, denoted by $\langle A_\alpha \rangle_{\alpha \in \mathbf{R}}$, with the property that $\alpha \neq \beta$ implies either $\alpha \in A_\beta$ or $\beta \in A_\alpha$. Then there is a sequence of functions $f_n \colon \mathbf{R} \to \mathbf{R}$ such that for any uncountable subset S of $\mathbf{R}$, we have $f_n(S) = \mathbf{R}$ for all but finitely many f_n.*

Proof: Using the existence of $\langle A_\alpha \rangle_{\alpha \in \mathbf{R}}$, we first construct a sequence of functions f_n with the property that for all α, and for all $\beta \in A_\alpha$, there exists an ordered pair (n, t) in the counterpart of β such that $f_n(\alpha) = t$. The construction is essentially a diagonalization technique. For each α, let the countable set A_α be enumerated as

$$\{\beta_1, \beta_2, \beta_3, \ldots\}.$$

Start with (n_1, t_1) being the first ordered pair in the counterpart of β_1. Proceed inductively, and let (n_k, t_k) be the first ordered pair in the counterpart of β_k such that $n_k > n_{k-1}$. This selection can be made because the first component of the counterpart of β_k is unbounded. Thus, we have constructed a sequence of ordered pairs $\langle (n_k, t_k) \rangle_{t \geq 1}$ with n_k increasing and each (n_k, t_k) in the counterpart of β_k. Using this sequence, we then define the function f_n by the rule

$$f_n(\alpha) = \begin{cases} t_k, & \text{if } n = n_k \text{ for some } k; \\ \alpha, & \text{otherwise.} \end{cases}$$

Indeed, f_n is well-defined since $n_i \neq n_j$ for $i \neq j$. Moreover, the sequence $\langle f_n \rangle$ has the desired property that for every α and every β in A_α, there is an ordered pair (n, t) in the counterpart of β such that $f_n(\alpha) = t$.

Now we show that any subset S of $\mathbf{R}$ for which infinitely many n have $f_n(S) \neq \mathbf{R}$ must be countable, thereby proving the theorem. If $f_n(S) \neq \mathbf{R}$ then there exists a real $t \notin f_n(S)$. So if there are infinitely many f_n such that $f_n(S) \neq \mathbf{R}$, then there is a sequence of ordered pairs (n, t) with n increasing and $t \notin f_n(S)$. Let the counterpart of this sequence of ordered pairs be β. Thus, every ordered pair (n, t) in the counterpart of β has $t \notin f_n(S)$. Now consider all real $\alpha \notin A_\beta \cup \{\beta\}$. By the hypothesis, we must have $\beta \in A_\alpha$. We constructed the sequence $\langle f_n \rangle$ in such a way that there is an ordered pair (n, t) in the counterpart of β with $f_n(\alpha) = t$. But by the choice of β, we have $t \notin f_n(S)$. Hence, $f_n(\alpha) = t \notin f_n(S)$ implies $\alpha \notin S$. Therefore S must be a subset of $A_\beta \cup \{\beta\}$, a countable set, implying that S is also a countable set.

Solution C

...If the real number α corresponds to the pair $(\langle n_k \rangle, \langle t_k \rangle)$, then we call $\langle n_k \rangle_{k \geq 1}$ the *integer sequence* of α and $\langle t_k \rangle_{k \geq 1}$ the *real sequence* of α.

...**Proof.** Note that a given real number α has associated with it both integer and real sequences, as well as a set of reals $A_\alpha = \{\alpha_1, \alpha_2, \alpha_3, \ldots\}$. We add to this list and construct an infinite set of integers $L_\alpha = \{l_1, l_2, l_3, \ldots\}$ in which each l_i comes from the integer sequence of α_i.

...

$$f_n(\alpha) = \begin{cases} t_n, & \text{if } n = l_i \in L_\alpha, \text{ where } \langle t_k \rangle \text{ is the real sequence of } \alpha_i; \\ \alpha & \text{otherwise.} \end{cases}$$

With these functions we will establish the contrapositive of the theorem: If $f_n(S) \neq \mathbf{R}$ for infinitely many integers n, then S is countable. ...

Solution D

As a step toward proving the Continuum Hypothesis, which states that there are no infinities between the countably infinite and the continuum, Sierpiński proposed the following theorem.

Suppose we have a function, $\text{spec}(\alpha)$, that maps every real α to a countably infinite subset of the reals (Figure A). Now suppose we make the additional hypothesis that for any two reals $\alpha \neq \overline{\alpha}$, either $\alpha \in \text{spec}(\overline{\alpha})$ or $\overline{\alpha} \in \text{spec}(\alpha)$ (Figure B). Then we can draw the following conclusion. There exists ...

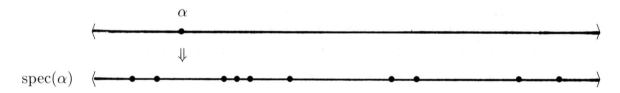

Figure A. Each real number α determines $\text{spec}(\alpha)$, a countably infinite subset of the reals.

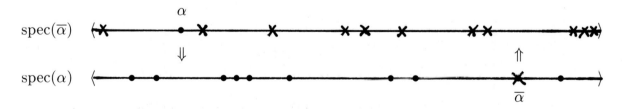

Figure B. By hypothesis, either $\alpha \in \text{spec}(\overline{\alpha})$, or $\overline{\alpha} \in \text{spec}(\alpha)$. Here α is not in $\text{spec}(\overline{\alpha})$, so $\overline{\alpha}$ must be $\text{spec}(\alpha)$.

§22. Excerpts from class, November 9 [notes by PMR]

> Quotation ... a writer expresses himself in quoting words that have been used before because they give his meaning better than he can give it himself, or because they are beautiful or witty, or because he expects them to touch a chord of association in his readers, or because he wishes to show that he is learned and well-read. Quotation due to the last motive is invariably ill-advised; the discerning reader detects it and is contemptuous, the undiscerning is perhaps impressed, but even then is at the same time repelled, pretentious quotation being the surest road to tedium.
>
> <div align="right">Fowler, <i>Dictionary of Modern English Usage</i>.</div>

> Mais malheur a' l'auteur qui veut toujours instruire! Le secret d'ennuyer est celui de tout dire.
>
> <div align="right">Voltaire, <i>De la Nature de l'Homme</i>.</div>

> Il ne faut jamais qu'un prince donne dans les détails. Il faut qu'il pense, et laisse et fasse agir: Il est l'aîme, et non pas le bras.
>
> <div align="right">Montesquieu, <i>Mes Pensées</i>.</div>

Don's secret delight, he confessed today, is to "play a library as if it were a musical instrument." Using the resources of a great library to solve a specific problem—now *that*, to him, is real living. One of his favourite ways to spend an afternoon is amongst the labyrinthine archives, pursuing obscure cross-references, tracking down ancient and neglected volumes, all in the hope of finding the perfect quotation with which to open or conclude a chapter. Don takes great pleasure in finding a really good aphorism with which to preface a piece of writing. So many people have written so many neat things down the ages, he said, that it behooves us to take every opportunity to pass them on. Don has been known to take such a liking to a phrase that he has written an article to publish along with it.

So how are we to find that wonderfully apposite quotation with which to preface our term paper? Serendipity, said Don. Live a full and varied life, read widely, keep your eyes and ears open, live long and prosper. You will stumble across great quotations. For example, Webster defines 'bit' as "a boring tool"—Don was able to use this when introducing a computer science talk.

Sometimes one needs to go about the search more systematically. For example, Don's TEXbook consists of 27 chapters, 10 appendices, and a preface. His format demands two relevant quotations at the end of each of these. His METAFONT book posed exactly the same problem. How did he go about it?

The first secret, he confided, is Bartlett. There are numerous dictionaries of quotations [filed under PN 6000 in the reference section of Green Library], of which Bartlett's *Familiar Quotations* is the most familiar. It was here, under the heading 'technique' in the index, that Don found a quote from Leonard Bacon deriding Technique as the death of true Art.

Now '$\tau\epsilon\chi$', in Greek, means both 'technique' and 'art', so this seemed pretty appropriate for *The TEXbook* where the (Greek) name TEX is explained.

When Bartlett fails, we can try the OED. This incomparable dictionary lists every word along with contexts in which it has been used; very often it prints a memorable quotation that incorporates the word in question. Likewise, we can turn to concordances of Shakespeare or Chaucer to find every single instance in which these authors used any given word.

Leafing through *The TEXbook*, Don picked out some of his favourites: Goethe on mathematicians (and why they are like Frenchmen); Paul Halmos telling us that the best notation is no notation (write mathematics as you would speak it!). Tacitus had something to say about the macro (or rather, about the ancient politician of that name).

A stiffer challenge was provided by a book that listed the METAFONT code defining each letter of the alphabet (as well as other symbols) in a certain typeface; Don had to come up with quotes for individual letters of the alphabet. No problem: James Thurber had proposed the abolition of 'O'; Ambrose Bierce had scathing things to say about 'M' in his famous *Devil's Dictionary*; Benjamin Franklin once wrote to Bodoni concerning the exact form of the letter 'T'; a technical report about statistical properties of the alphabet deliberately made no use of the letter 'E'.

Some of the best quotations are taken entirely out of context. The economist Leontief had something to say about (economic) output; Don quoted him in his chapter on (computer) output. Galsworthy's comments on Expressionists found their way into his section on expressions.

In a pinch, said Don, quote yourself. You could even find someone famous and ask her to say something—anything!—on such-and-such a subject. In another desperate case, Don couldn't find anything much that had been said about fonts. No matter, he quoted the explorer Pedro Font writing about something else entirely (the discovery of Palo Alto, as it happens). If you are Don Knuth, you may even be able to quote Mary-Claire van Leunen praising your use of quotation!

Computer technology now gives us another quote-locating resource. When Albert Camus' *The Plague* is available online, it will be a simple matter for this note-taker to find the part in which a writer agonizes for a week before putting a comma in a particular sentence, and then for another week before taking it out again; just search for occurrences of the word 'comma' in the text. Don used this technique to find quotations involving the word 'expression' in *Grimm's Fairy Tales* and *Wuthering Heights*, both of which are available on SAIL.

If any member of the class would like to demonstrate virtuosity at "playing the library," he could try to track down the quotation "God is in the details." Don rather identifies with God in this, but hasn't been able to track down the reference. A number of people have assured him that it originated with Mies van der Rohe, but despite reading all the works and contacting the two biographers of this architect, he has not been able to find it. Someone told him that Flaubert once wrote "Le bon Dieu est dans le détail." Don hasn't the patience to search exhaustively in Flaubert's voluminous publications, but he did try

French equivalents of Bartlett—finding the two quotes above (which express the opposite sentiment). The God-in-details aphorism remains an orphan to this day.

Don has found another quote that so well expresses his philosophy on the subject of error that he is having it carved in slate by English stonecutters, to occupy pride of place in his garden:

>> The road to wisdom? Well it's plain
>> and simple to express:
>>> err
>>> and err
>>> and err again
>>> but less
>>> and less
>>> and less.

Mention was also made of indexes for books. The Sears & Roebuck catalogue for 1897 contains the useful advice: "If you don't find it in the index, look very carefully through the entire catalogue." A British judge named Lord Campbell wanted legislation to compel writers to index their work, but was unable to get round to indexing his own.

Tangentially, Don mentioned that the designers had given his TEXbook rather large paragraph indentations—perhaps it's the style of the 80s, he said. This meant that he sometimes had to add or subtract words to ensure that the last line of each paragraph was at least as long as the indentation on the following one. The page looks rather strange if this isn't the case.

§23. Excerpts from class, November 11 [notes by TLL]

Today we heard war stories—stories of the wars between Don Knuth and the *Scientific American* editorial staff.

However, before we got completely on track, Don told us a little about the the book he is writing this quarter: *Concrete Math*.

He said that this summer he went to see *Snow White and the Seven Dwarfs* and was very impressed. ("Who would have conceived, in 1937, that such a work of art could be made?") He said he was inspired; that he wanted to produce a work of art as inspired as *Snow White*, "except that I wanted to finish it in three months."

A book in three months: This means that Don has to "crank out" four pages a day, including Saturdays, Sundays, and Holidays. Surprisingly, Don says, "Here it is November, and I am still happy." He says sometimes he gets up in the morning and can't wait to get writing; at other times he just finds it a chore that he has to do; "but once I get started, it's easy—starting is the hard part."

At this point he delivered the punch line to his story on inspiration: We have one more week to finish the first draft of our term papers. We have the good fortune to have two professional editors who have volunteered to read our papers: Mary-Claire van Leunen and Rosalie Stemer.

Moving immediately from his statement that we were lucky to have professionals editing our work to the stories of his wars with a professional editor, Don showed us a quotation from *The Plague*, by Camus [found by PMR].

> "What I really want, doctor, is this. On the day when the manuscript reaches the publisher, I want him to stand up—after he's read it through, of course—and say to his staff: 'Gentlemen, hats off!'"

Of course, the fictional character who made the above statement is portrayed by Camus as being not only naïve but a bit mentally unstable. This doesn't mean that a person couldn't harbor a healthy enmity for an overzealous copy editor.

We now review the correspondence concerning one paper that Don eventually had published in *Scientific American* (henceforth known as SA):

In the Fall of 1975, Don received a letter from Dennis Flanagan (the editor). The letter invited him to write a paper, of about 6000 words, on the topic of Algorithms, for SA's 600,000 readers. It offered him a $500 honorarium for such a paper. (This means he got about eight cents per word or eight cents per 100 readers—depending on how you like to think of such things.)

After some correspondence concerning the date that the paper was to be received (Don had been ill and the date needed to be pushed back), we came to the cover letter for the original manuscript that Don submitted to SA. He told Mr. Flanagan that he understood that some editing would take place, but that he had gone out of his way to try to imitate the "*Scientific American* style." Don told them, "It will be interesting to see what you do to this, my masterpiece."

Don soon got a letter back from Mr. Flanagan acknowledging Don's paper, telling him that it might have to be "slightly edited," and warning him that it might take a while to give it the attention it deserves. (Don also got his $500 at this point.)

Finally, 14 months later, Don received an edited copy of his paper together with a cover letter that explained that it had been "edited for the general reader." Don was told to correct any errors that they might have inadvertently introduced and exhorted to get back to them within the next two weeks.

To put it mildly, Don was not pleased with the results of this editing. Every sentence had been rewritten. He wrote a letter to Martin Gardner—a letter written more to vent frustration than in expectation of achieving any result—in which he stated many of his grievances. One of his comments covers the general tone: "I was astonished to see how many editorial changes were made that took perfectly good English and turned it into something that would be worth no more than B^- on a high school term paper."

In addition to showing us his letter to Gardner (and Mr. Gardner's sympathetic response) he showed the class the original and the edited versions. Among SA's changes: Changing all uses of 'we', transforming some long sentences to several short sentences, transforming some short sentences into one long sentence, removing commas (commas that Don found necessary), changing 'which's to 'that's, removing technical jargon, changing 'most common' to 'commonest', and introducing a few errors. (Don found many of the changes gratuitous, but the editorial introduction of errors was useful because it meant that Don's exposition had not been clear enough.)

The next letter we saw was the cover letter for the, now re-edited, manuscript that Don sent back to Dennis Flanagan. He mentioned his extensive re-editing, stated that he appreciated some aspects of the editing more than some others, and asked to see the galley proofs; he said viewing the proofs was especially important since there is "so much technical material that is typographical in nature."

Two weeks later Don got back the proofs and a letter. The letter argued successfully with some of Don's objections to the original editing job (they stuck by 'that' instead of 'which', hurray!); less successfully, SA refused to budge on 'commonest' (boo). The letter also said that sending proofs to an author was unprecedented. But the printer was having a terrible time with the mathematics, so they made an exception. (Don pointed out that this was largely dictated by the printing mechanisms they were using.) But it was a good thing the proofs were sent, because important changes were made during a 1.5-hour telephone conversation.

By the end of class, about the time that Don showed us his second letter to Martin Gardner—the one in which he said he shouldn't have been so frustrated in the first place—Don admitted that some of the disputed changes really had been appropriate ones. He said that the original copy editor had improved his article in some ways, but that his further editing had improved it still further. At the end everybody was happy. (Music up.)

As a final parenthetical remark, he told us about the way that the (quite long) captions for illustrations in SA are typeset: The linebreaks are determined by hand. The final line always ends at the right margin (there's no extra white space). To achieve this, the SA copy editors must count letters and reword the captions until they fit. One of the methods that they use to make things fit nicely is to start at the end of the caption and start removing 'the's. ("It is not placed at root of tree because it is too far from center of alphabet.") At least, this was the system in 1977.

Among comments about how the SA editorial staff is overworked and how he shouldn't have been so upset, he did get off a parting shot: "After spending all this time doing crazy stuff like caption filling, it's no wonder the copy editor had no time for polishing my article."

§24.   Friday the 13th, part 24: The Classnotes   [notes by PMR]

The Story So Far. *Readers will recall that our hero, 'Prof' Don, is locked in mortal combat with Scientific American, a journal whose global reach is exceeded only by its editorial hubris. Will Don's definitive Algorithms article reach the world unscathed? Or will it suffer the death of a thousand 'improvements' at the hands of a hoard of dyslexic copy-editors?* **Now Read On . . .**

On March 25, Don received the page proofs for his article, which was to appear in the April edition. ("Ever since Martin Gardner's famous April Fool hoax, I had wanted to get into an April issue," he mused.) Don picked up the phone and spent the next hour and a half in damage-limitation negotiations with an editor code-named **TEB**.

Some straightforward errors were easily corrected: A '1' had metamorphosed into an 'l' and an '∅' into a 'φ'. Typesetters who are unfamiliar with mathematics invariably find creative things to do with this "empty set" symbol, Don said. Many problems show up only at the page-proof stage. For example, one page began with the solitary last line of a paragraph and then broke with a new subheading. Since the paragraph could just as easily introduce the new subsection as conclude the previous one, Don just moved the subheading back one paragraph, putting it on the previous page. Don also got his floor brackets restored where square brackets appeared on the page proof.

He didn't get his way on everything, though. Brackets were used interchangeably with parentheses in a mathematical formula, despite Don's protest that the former have special meanings.

Neither was *Scientific American* ('SA', hereinafter) able to get hold of a photograph of a particular Mesopotamian clay tablet that is housed in the Louvre. It is a table of reciprocals, and is probably the earliest example of a large database that was sorted into order for ease of retrieval. Don thinks this object definitely deserves a place in the hearts and minds of CS folk, being perhaps the first ever significant piece of data processing. Even a modern computer might need a second or so to do the work involved.

On the whole, Don was pretty happy with his article. It enjoys a continuing success as an SA reprint; thousands of copies are still sold to schools (with the page references carefully renumbered). As far as Don knows, it's the only one of his articles to have been translated into Farsi (Persian). He showed us that in this language, as in others where the text runs right to left across the page, mathematical formulæ are not reversed. The word 'hashing' invariably gives translators pause; it becomes 14 characters in Chinese, and a French translator of one of his books once put in a call to the Academie Française to establish the authorized equivalent.

All the re-editing was painful at the time, admits Don, but in the long run he has come to agree that this coöperative effort did much to remove the jargon and make the paper accessible to a general audience. Martin Gardner, Don told us, attributes his success as a mathematics writer to the fact that he is not a mathematician.

Don's paper for the IEEE *Transactions on Information Theory* makes for a sadder tale: They made such a mess of it that Don decided the game just not worth the candle, and he advises everyone to read the Stanford CSD Report instead. For example, IEEE says 'zero' and 'one' instead of '0' and '1'. Don likes to use 'lg' to mean 'log to the base 2', but they changed this without explanation to 'log' despite the fact that to most people this latter means 'log to the base 10'; or to number theorists, when it means the natural log (base e). Not the greatest copy-editors, Don sighed.

More recently, Don wrote for the October issue of the ACM *Transactions on Graphics*, and encountered some really shocking copy-editing. They changed '...data has to ...' to '...data have to ...'. Now long ago Don was told that 'data' is *really* plural, but everywhere it is used both as a singular or a plural, even in the reliably conservative ('antediluvian!' chimed Mary-Claire) *New York Times*. Don thought it quite right to use it as a singular when referring to data as some kind of collective stuff. Don wrote and complained that the ACM should certainly know about data. In the end, Don kept everyone happy by changing the sentence to read '...data must...'.

Mary-Claire van Leunen sanctioned the term 'Automata Theory', although one would not normally incorporate a plural adjective into a compound noun. But no-one has ever said 'automaton theory', and no-one ever will.

ACM did gracefully admit to and correct some straightforward mistakes, such as 'this number plus that number are equal to 63'. But where Don wrote 1000000 they substituted 1,000,000. Don objected that although this might be justified in text, his use is perfectly OK in a *formula*. Well then, they replied, write 10^6. Fine, said, Don, but what do I do when the number is 1234567? The IEEE standard here is to insert spaces, thus: 1 234 567. Don doesn't like this in formulæ, but agrees that it may be useful in a high-precision context, such as numerical tables.

Don recalled a remark by George Forsythe that every scientist should try to write for a general audience—not just for other scientists—at least once in his life. Don has done this three times now, so feels that he's done his bit! He gave his first such lecture to a non-technical audience in Norway and found it surprisingly hard to understand their 'mind set'. The problem is to make the talk interesting, but convey how it feels to a computer scientist to do computer science. The public probably imagine that mathematicians sit and factor polynomials all day, and that CS types design videogames. How to convey the soul of the subject to them? In this lecture, Don presented a sequence of algorithms for a search task. Since we all have to look up information in large tables or indexes now and then, he hoped the audience would have a clear intuition of the problem. Brute force searching is clearly too slow; binary search is natural and powerful; hashing is better still, but very unintuitive to most people. Don was asked to write up his talk for a Norwegian magazine called *Forskningsnytt*, 'Research News' (a sort of *Scientific Norwegian*). In the course of doing so he learned enough of the language to write v and h instead of l and r to designate left and right sons in a tree structure. Dr. Ole Amble, a numerical analyst who was one of Norway's computer pioneers, helped Don with Norwegian style on this article, and got interested in search algorithms as a result. He asked Don whether there mightn't be a way to combine the advantages of binary search and hashing? Don at first told him "obviously not," but then realized what Amble meant ...alas, too late to include in the just-published Volume 3 of *ACP*. But this combination of methods made a nice conclusion to his SA paper, which was based on this Norwegian prototype.

It was in April of 1977 that Don's travails with SA prompted him to investigate typesetting for himself; in May of that year he designed the first draft of TeX and spent his sabbatical (and ten more years) perfecting it, putting Volume 4 of *ACP* on the back burner.

We had a few minutes left to look at other changes that SA made to Don's original manuscript. In one case, there seemed to be no reason for restructuring a sentence to put Amble's name first instead of the motivation of his discovery. But Mary-Claire noted that SA always tries to stress the human contributions in science, sometimes at the expense of the ideas. Don also mentioned another surprising thing he learned about SA's editorial policy: They never display equations. (PMR knows at least one scientist who refuses to read SA for this very reason—"How can you explain science without equations?—Pah!")

§25. Excerpts from class, November 16 [notes by TLL]

After a brief (but charming) musical prelude, Don demonstrated to us that we are not alone in being concerned with the mechanics of writing. He showed us four small publications that touched on some of the humorous aspects of written rhetoric.

We briefly viewed a Russell Baker column entitled "Block That That Cursor"; a "Peanuts" comic strip with a punchline concerning comma placement; a quotation from the *New York Times* ("Plagiarize creatively, but quotes can be dangerous if you don't acknowledge the source"); and an article by Richard Feynman in the Caltech *Alumni* magazine. Feynman discussed his disappointment with his experience of serving on the Challenger Disaster Panel; he complained that instead of discussing ideas, the panel spent all their time "word-smithing" (deciding how to reword or re-punctuate sentences in the committee's report).

Feynman's dismay at the amount of time he spent dealing with commas, wicked-whiches, and typographic presentation is not unique. Don said, "Word-smithing is a much greater percentage of what I am supposed to be doing in my life than I would have ever thought. That's one of the main reasons I am teaching this course."

Don also showed us what he thinks is a wonderful piece of writing: a spoof on the Sam Spade genre, full of detectives, blondes, .38's, and the 'sweet smell of greenbacks'. It turned out to be a passage from "Getting Even" by Woody Allen. Likewise for the term papers, he said, try to have a genre in mind (though perhaps not this one) and do a good job in that genre.

To help prepare us for the guest speakers coming up soon, Don handed out copies of several of their works, encouraging us to read them as examples of good practice. First he handed out an "Editor's Corner" article published by Herb Wilf last January:

> This issue marks another changing-of-the-guard for the MONTHLY. Paul Halmos' act will be a tough one to follow ...

Wilf's article contains a nice exposition of problems related to Riemann's famous unproved Hypothesis.

Don also showed us another draft of a paper by Herb: "n coins in a fountain". This title, he said, was just too good to pass up, even though it includes a formula. But Don would have capitalized the n, because it comes first. As for the objection about starting a title with a symbol, why shouldn't we regard N as simply another English word? (After all, it appears in most dictionaries as the first entry under 'N'.) But this approach would make it necessary to capitalize N throughout the article.*

The next guest speaker after Wilf will be Jeff Ullman, who will tell us how to become rich by writing textbooks. Don recommended that we look closely at Chapter 11 of Jeff's book *Principles of Database Systems* (second edition), which shows "excellent simplification of subtle problems and algorithms."

---

* Herb eventually solved the problem by calling his paper 'The Editor's Corner: n Coins in a Fountain', in *American Mathematical Monthly* **95** (1988), 840–843.

Don handed out two examples by the third guest speaker Leslie Lamport. One, from *Notices of the American Mathematical Society* **34** (June 1987) is entitled "Document Production: Visual or Logical?" and Don said "It's a 'flame' but very well written so I wanted you all to read it. It's a nice polemic that takes the 'WYSIWYG versus Markup' controversy and reformulates the problem along more fruitful lines." The other Lamport article is entitled "A simple approach to specifying concurrent systems"; it will soon be published in *Communications of the ACM*.

Don says the latter paper is the best technical report he has seen in the last year or so. The paper is unusual because of its question-and-answer format. While dialogs have been used effectively by experts in other fields (such as Socrates, Galileo, George Dantzig, and Alfred Rényi), this is the first time, as far as Don knows, that such a format has been used in computer science.

> Before moving on to the next handout, Don told us about writing his book *Surreal Numbers*. Like Leslie Lamport's paper, Don's book is presented as a dialog. Don's dialog presents some ideas that John Conway told him at lunch one day (Don wrote the ideas down on a napkin and then lost the napkin). The most extraordinary aspect of this book is that Don wrote it in six days ("And then I rested"). That week was very special for Don. ("It was the most exciting week in my life. I don't think I can ever recapture it.")
>
> When Don wrote the book he was in Norway. He was in the middle of writing one of the volumes of *The Art of Computer Programming* (isn't he always?), and he did not expect Jill (his wife) to be sympathetic when he told her that he wanted to write yet another book—even if he did think he could write it in a week. Perhaps Jill knows more about Don than Don knows about Jill, because she not only didn't complain but she got quite into the spirit of the thing.
>
> Just what was the spirit of the thing? "Intellectual whimsey" probably isn't far off. Don rented a hotel room ("near where Ibsen wrote") and spent his week writing, taking long walks ("to get my head clear"), eavesdropping on his fellow hotel guests at breakfast ("so I could hear what dialog really sounds like"), and pretending that Jill's visits were clandestine ("we had always read about people having affairs in hotels …").
>
> Don said he wrote "with a muse on my shoulder." Every night's sleep was filled with ideas and solutions; before dozing off he would have to get up and write down the first letter of every word of the ideas he had (and he would spend the morning decoding these cryptic scribbles). He told us that he was more perceptive during this week— his description of the King's Garden during an evening walk was worthy of Timothy Leary.
>
> All this prolific word production must have left him in verbal debt: When he finished the book he tried to write a letter to Phyllis telling her how to type the book. He couldn't. Except he must have eventually—the book is still in print and sells several hundred copies a year (in seven languages).

Still another handout was part of a chapter written by Nils Nilsson and Mike Genesereth for their new book *Logical Foundations of Artificial Intelligence*. Chapter 6, entitled "Nonmonotonic reasoning," presents a new area of research at the level of a graduate student. Don says that the chapter has an excellent blend of formal and informal discussion, with well-chosen examples; this subject had never been "popularized" before, so the task of writing a good exposition was especially challenging. Don also praised the authors' typographic conventions (for example, logic is presented using a "typewriter" font).

Don said that we already have Mary-Claire's book, so he didn't have to introduce her to us. But he ran across some electronic mail she had written recently, and thought it was a particularly elegant essay, so he passed it along (see §26 below). Computer scientists and mathematicians are way behind real writers when it comes to exquisite style.

Finally, just in case we still craved more good examples to read, he handed out some excerpts from a paper written by Garey, Graham, Johnson, and Knuth. Don says that he included it because it has two proofs of difficult theorems: proofs that are not, and probably could not be, trivial.

Don tried to interest his readers in the first proof (and algorithm) by presenting an example as a mathematical puzzle. He says that by solving the puzzle the reader can see that the problem is not simplistic—but that an algorithm might be possible. ("This builds exactly the right mental structures in the reader's mind for this particular problem, I think. The algorithm itself is the worst algorithm I have ever had to present—but there is probably no simpler one.") While flashing us part of the algorithm—complete with more cases than could fit on the monitor—Don said, "The ability to handle lots of cases is Computer Science's strength and weakness. We are good at dealing with such complexity, but we sometimes don't try for unity when there is unity."

The second proof involves the reduction of one problem to another. The reduction requires a very complicated system—a system that Don found was well served by an extended biological metaphor and some involved terminology. As his metaphor, he chose the jellyfish ("an unrooted, free-floating tree"); he named pieces of the data structure stems, polyps, tentacles, heads, and nematocysts (the biological term for stingers).

Mary-Claire asked, "If that structure turns out to be generally useful, are you going to be sad that you called it a nematocyst rather than a stinger?" Don said No, but he has been sorry about names he has chosen in the past. (He wishes he had called LR(k) grammars L(k) grammars.) When he was writing *The Art of Computer Programming*, Volume three, he used the word "Daemon" to refer to what are now called "Oracles," but the Oracle replaced the Daemon before it was too late.

Another last minute terminology substitution happened when Aho, Hopcroft, and Ullman substituted "NP-complete" for "Polynomially-complete" in their text on Algorithms—even though they had already gotten galley proofs using the original name. The name was changed at that late date as the result of a poll conducted throughout the Theoretical Computer Science community (suggested names were NP-Hard, Herculean Problem, and Augean Problem).

§26.

```
------ Forwarded Message

Replied: Forwarded 14 Aug 85 12:11
Return-Path: <mcvl>
Received: by lewis.ARPA (4.22.01/4.7.34)
         id AA20294; Mon, 6 Aug 84 12:22:47 pdt
From: mcvl (Mary-Claire van Leunen)
Message-Id: <8408061922.AA20294@lewis.ARPA>
Date:  6 Aug 1984 1222-PDT (Monday)
To: adams, ar.nrt@Stanford.BITNET, asente@Shasta, baldwin@Yale.ARPA, baskett,
        beigel, bell, bent@Wisc-rsch.ARPA, blatt, cnelson, decvax!jmcg,
        ellis@Yale.ARPA, estrin@MIT-XX.ARPA, feir, gnelson, goodman@Yale.ARPA,
        guarino, herbison@ultra.DEC, heubert@Yale.ARPA, horning, hsu@erlang.DEC,
        johnsson, karlton, kelsey, kmc, larrabee, levin, li, lowney@Yale.ARPA,
        mbrown, mccall.pa@Xerox.ARPA, mcdaniel, mcvl-essays@Purdue.ARPA,
        mcvl-essays@Washington.ARPA, meehan@Yale.ARPA, minow@rex.DEC,
        naughton, parker@Yale.ARPA, peters, petit, philbin, pierce, ramshaw,
        LEICHTERJ@RANI.DEC, rees, reid@Glacier, rentsch@unc.CSNET, ritter@Yale.ARPA,
        robson.pa@Xerox.ARPA, ruttenberg@Yale.ARPA, shivers, siddall, smokey,
        so.pa@Xerox.ARPA, stewart, swart, trow@Xerox, wall, wick, wilhelm,
        wittenberg@Yale.ARPA, wli%Yale-Ring@Yale.ARPA, young@Yale.ARPA
Cc: mcvl
Subject: #5: "hopefully"
```

Q: I have often heard that it is incorrect to use "hopefully" to mean "it is hoped." But the Random House dictionary lists that as definition number two and gives no warnings of any kind. Is this usage standard?

A:

The story on "hopefully" is one of the strangest in modern English.

"Hopeful" has had two senses ever since it first appeared in the language late in the 16th century. A person could be hopeful (expectant, eager, desirous); or a situation could be hopeful (promising, auspicious, bright).

As with most adjectives, both of these "hopeful"s regularly produced "-ly" adverbial forms, but the kind of hopefulness that means expectant and eager produced adverbs more readily than the kind that means promising and bright. There's nothing mysterious about that difference in frequency. A person can carry himself hopefully or eye a desirable object hopefully or prepare himself hopefully for a possible future. Impersonal substantives, on the other hand, serve less often than personal ones at the head of the kind of active verbs we modify with adverbs of manner. Nonetheless, a wager can be shaping up hopefully, a day can begin hopefully, the omens can augur hopefully. All perfectly straightforward and normal. The first OED citation for "hopefully" in this second sense is from 1637.

Early in the 1930s, this second sense of "hopefully" began to appear in a different kind of construction, as what's called a sentence adverb. Sentence adverbs are part of a class of expressions that can modify whole clauses; such expressions are called absolutes. Look at some sentence adverbs at work:

> Interestingly, most mathematicians failed to notice the correspondence.
>
> Presumably he knows what he's doing.

> Regrettably there is no remedy for this kind of infection.
>
> Fortunately we managed to get out before he noticed us.
>
> Hopefully the weather will clear up before it's time to leave.

Then in the early 1960s attacks against "hopefully" began to appear in print. That's about the right lag time for usage controversies, and I looked forward to figuring out how to wield my cudgel. I believe, by the way, that if rational debate had ensued I would have been against "hopefully" as a sentence adverb. Unfortunately, reason never entered into it.

The attacks were by and large astoundingly ill-informed. Some managed to convey that it was "hopefully" as a sentence adverb to which they were opposed but lacked the technical vocabulary with which to express the idea. Some opposed both "hopeful" and "hopefully" in the sense promising(ly), auspicious(ly), bright(ly). Most bizarre of all, some took it upon themselves to oppose all sentence adverbs.

The authors of these attacks presented themselves as defenders of the purity of the language against the onslaught of wicked barbarians. They asserted (apparently without ever feeling the need to check the evidence) that the objects of their attack were very recent additions to the language -- true in the case of "hopefully" as a sentence adverb, but not true at all of impersonal "hopeful" and "hopefully," and completely zany when it comes to sentence adverbs in general. (In five minutes with the OED I was able to find "certainly" being used as a sentence adverb in 1300.) One came to see that these self-proclaimed language defenders knew nothing of even the most elementary tools of the trade.

So. What should we do? Ignore the ignorant bully-boys or knuckle under?

When I was young I once attended a school at which one group of girls declared that anyone who wore yellow on Thursday was a freak. The rest of us recognized the interdict as arbitrary, irrational, and mean. Its only purpose was to wound. We talked about it among ourselves, and we tried to fire ourselves up to wear yellow on the fatal day. But there were sixteen lonely hours between last study hall on Wednesday and getting dressed Thursday morning. Nobody owned very many items of yellow clothing anyway; quite likely they were already in the laundry. And yellow's not a flattering color in the morning light. Not at all flattering. Tends to make the wearer look green.

I moved in the middle of the year, so I have no idea whether some brave child eventually wore a yellow blouse, or yellow socks, or a yellow handkerchief to that school on a Thursday. Hopefully there was more than one; hopefully the girls who laid down the original rule saw how jaunty the rebels looked in their yellow outfits; hopefully -- oh, devoutly to be hoped --- they all became best friends and behaved beautifully to one another and never did anything petty or malicious again as long as they lived.

§27. Excerpts from class, October 28 [notes by TLL]

Class opened as Don introduced today's guest speaker: Professor Herbert Wilf. Professor Wilf is on the faculty at the University of Pennsylvania but is spending his sabbatical year at Stanford.

As Wilf took the dais he pronounced this "a marvelous course." ("Taken earlier in my career it would have saved me and the world a lot of grief—mostly me.") The course topic is one of daily concern for him; apart from writing his own papers he edits two very different journals: the *American Mathematical Monthly* ("The MONTHLY") and the *Journal of Algorithms*.

The *Journal of Algorithms* was founded in 1980 by Wilf and Knuth and is a research Journal. Results are reported there if they are new, if they are important, and if they are significant contributions to the field. If these conditions are met, a little leeway can be given in the area of beautiful presentation. But the MONTHLY is an expository journal. It is a home for excellent mathematical exposition. (It also seems to be a popular place to send "proofs" of Fermat's Last Theorem.)

Though he told us that he feels "older without feeling wiser" and is uncomfortable setting down rules for a human interaction that "involves part brain and part hormone system" he gave us several pointers.

> **Get the attention of your readers immediately.** Snappy titles, arresting first sentences, and lucid initial paragraphs are all methods of doing this.
>
> As examples, he showed us a paper by Andrew M. Gleason with the title "Trisecting the Angle, the Heptagon, and the Triskaidecagon"; a paper by Hugh Thurston that began "Can a graph be continuous and discontinuous?"; and the first paragraph of an autobiographical piece by by Olga Taussky-Todd that started with some insight into the author's fascination with matrices. Gleason's paper was attention-getting mostly because Gleason is famous—"that helps."
>
> **Get everything up front.** Tell your readers in plain English what you are going to write about and let them decide for themselves whether or not they are interested. ("You can quintuple your readership if you will let them in on what it is that you are doing.")
>
> **Remember that people scan papers when they read them.** Potential readers will skim looking for statements of theorems; if all of your text is discursive they will have nothing to latch onto. Summarize your results using bold face ("or neon") so that the page flippers can make an informed decision. Similarly, drop notational abbreviations and convoluted references in the statements of theorems.
>
> **A little motivation is good, but readers don't like too much.** Presenting examples that do not yield desired results can be quite useful, but the technique loses its charm after a small number of such examples. (Far from overdoing this technique, many writers will introduce mysteriously convenient starting points for

their theorems. "Whenever I see 'Consider the following ...' I know the author really means to say 'Here comes something from the left field bleachers.'"

He gave us the name of three books (not written by anyone in the room) that he considers superb books of mathematics:

Problems and Theorems of Analysis, by Pólya and Szegő. It has a "Problems" section and an "Answers" section. The problems are self-contained, digestible pieces of more complex problems. The answers are on the spare side and have been the cause of much head-scratching over the years. By solving several of these self-contained problems, a reader can arrive at an understanding of major results in the field.

Introduction to The Theory of Numbers, by Hardy and Wright. This book is "short on motivation." Theorems are stated and proved concisely and precisely. In the preface the authors claim that "the subject matter is so attractive that only extravagant incompetence could make it dull."

Mathematical Analysis, by Rudin. This book is rigorous. It teaches the reader what is and is not a proof. A reader who survives this book feels strong.

Wilf commented that all three of these books are quite dry, but Knuth objected (along the same lines as those used by Hardy and Wright in their preface) and Wilf amended his statement: Each of these books is very lean.

Discussing the change of his own writing style over time, he told us that when he was younger he didn't have much self esteem and stuck to established forms. Now that he feels better about himself he has developed his own, much chattier, style. (Speaking of chattiness, he is also a fan of the use of the first-person in technical writing.) He says he aims to be chatty leading up to a proof, prove it in the "lean and mean" style that Rudin would use, and then be chatty again after he finishes the proof.

The last things that Wilf discussed were two handouts (§28 and §29 below): "Enumeration of orbits of mappings under action of C_n, the cyclic group," and "Counting necklaces." Each handout discusses the same mathematical problem, solved the same way. "Enumeration of ..." takes a half page; "Counting Necklaces" takes four pages.

Some audience members will appreciate the half page of exposition that is condensed to the word "evidently" in the shorter paper; some will merely be annoyed by it. As the *Monthly* editor he gets letters from people who complain about the informal style creeping into recent publications. "Mathematics is a serious business, not a comic pursuit," said one such letter.

Finally, Wilf doesn't mean to say that either of the two approaches is superior ("They are the two sides of the coin"); he means for us to examine each and decide what techniques we want from each.

§28. From Acta Hypermathica

Enumeration of orbits of mappings under action of C_n, the cyclic group

B. Nimble

Say that a mapping $f : [n] \to [k]$ is *irreducible* if $\forall \{\sigma \in C_n; \sigma \neq 1\}$ we have $f \circ \sigma \neq f$. If $M(n,k)$ is the number of these and $F(n,k)$ is the number of orbits of mappings f under the action of C_n, then evidently

$$F(n,k) = \sum_{d|n} M(d,k) \quad (n \geq 1). \tag{1}$$

But since, clearly,

$$\sum_{d|n} d M(d,k) = k^n \quad (n \geq 1) \tag{2}$$

we find from (2),

$$M(n,k) = (1/n) \sum_{d|n} \mu(n/d) k^d$$

and from (1),

$$F(n,k) = \sum_{d|n} (1/d) \sum_{\delta|d} \mu(d/\delta) k^\delta$$

$$= (1/n) \sum_{d|n} \phi(n/d) k^d$$

where ϕ is Euler's function, and the last equality follows from simple manipulations.

§29. From MathWorld — Counting necklaces

R. U. Certain

Suppose we have a supply of beads of k different colors, and we wish to construct necklaces of n beads. How many different necklaces can we make? The word 'different' is to be understood in the sense of rotations; two necklaces are equivalent if one can be carried into the other by a rotation. Another interesting problem would have resulted if we had allowed the customer to pick up the necklace off of the counter and flip it over. In the latter case we would have been studying equivalences under the action of the *dihedral* group (generated by a cyclic shift right by 1 unit and a 180° flip) instead of under the *cyclic* group, which is in fact what we're going to talk about here.

For instance, if $n = 4$ and $k = 2$ there are 6 different necklaces, and these are shown below.

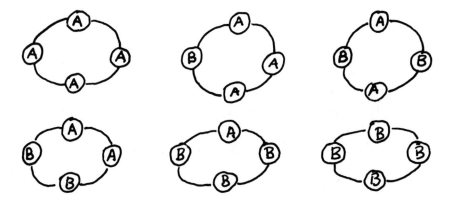

The problem is to find $F(n, k)$, the number of different n-bead necklaces of at most k colors (let's call these (n, k) necklaces).

Among all (n, k) necklaces we distinguish a subset that we will call the 'prime' necklaces. Say that a necklace is prime if it does *not* result from concatenating a number of repetitions of a shorter pattern. Thus, among the $(4, 2)$ necklaces above, the first one is not prime because it results from stringing together 4 identical shorter strings (viz., 'A'). The third and sixth ones are also not prime, whereas the second, fourth and fifth ones are prime.

Let $M(n, k)$ denote the number of *prime* (n, k) necklaces (e.g., $M(4, 2) = 3$).

The reason for concentrating on these prime necklaces will now appear.

Construction. *Begin with some divisor d of n and a prime (d, k) necklace. Cut the necklace immediately to the right of one of its beads. This yields a linear string of length d. Make n/d copies of that linear string and concatenate them to produce a single linear string of length n.*

We claim that *every possible one of the k^n possible linear strings of n beads of k colors can be constructed once and only once by such a cutting operation.* To prove that, let w be such an n-string, and let d be the smallest integer with the property that w is the concatenation of n/d copies of a string of length d. There is always at least one candidate for such an integer d, since $d = n$ will work. Evidently, whatever d is, it is certainly a divisor of n.

Having found d for the given string w, construct a necklace of d beads by taking the first d beads from w and tying their ends together. The resulting (d, k) necklace is *prime* (why?) and it is the one and only prime necklace with the property that if we apply the 'Construction' stated above, the given string w results.

Hence every prime (d, k) necklace yields d different linear strings, and every linear string turns up at some point in the game. That is to say,

$$\sum_{d|n} d M(d, k) = k^n \quad (n \geq 1).$$

After Möbius inversion, we get

$$M(n, k) = (1/n) \sum_{d|n} \mu(n/d) k^d \quad (n \geq 1), \tag{1}$$

where μ is the Möbius function. Hence we have an explicit formula for the number of prime necklaces.

Unfortunately, that wasn't the question. The question was to find the number of *all* different necklaces whether prime or not.

Fortunately, the latter number, $F(n, k)$, is easily obtainable from $M(n, k)$. Take a divisor d of n and a prime (d, k) necklace. Cut it somewhere, and make n/d copies of the resulting string. Concatenate them (sound familiar?) into a single n string, but now (this part shouldn't sound familiar) tie its ends together. The result of this operation is a *necklace*, not just a *linear string*. Hence it doesn't matter where we make the cut: wherever the cut is made, the end result is the same necklace.

Bottom line: $F(n, k) = \sum_{d|n} M(d, k)$

Since we have an explicit formula (1) for M and an explicit formula for F in terms of M, we're finished, aren't we? Well in a sense yes, but if we stick with it we'll find that simplifying the expression is half the fun. Where we are is that

$$F(n, k) = \sum_{d|n} M(d, k)$$
$$= \sum_{d|n} (1/d) \sum_{d'|d} \mu(d/d') k^{d'}.$$

The first step in the simplification process is to invoke the Law of Double Sums: 'Interchange Them'. This gives

$$F(n, k) = \sum_{d'|n} k^{d'} \sum_{d'|d|n} \mu(d/d')/d, \tag{2}$$

in which the innermost sum is over d's that are simultaneously divisors of n and multiples of d'.

Now that kind of a sum is very confusing to handle unless you use a language that has been attributed by a well-known SU professor to Iverson, although said professor has since developed its use to a fare-thee-well. The idea is to move all of the fine print up from the bottom of the airline ads into the main text so you can see that you have to stay over a weekend to get thee best fares.

The way to apply this sage advice (thymely too) is to use the function $T(\cdot)$, the 'truth-value' function. Its value on any particular '$\cdot$' is 1 if its argument is true and 0 else. In this case we have,

$$\sum_{d'|d|n} \mu(d/d')/d = \sum_d T(d'|d)T(d|n)\mu(d/d')(1/d).$$

But $T(d'|d) = T(\exists t : d = td')$, so we can replace d by td' in the inner sum above, and sum on t. The inner sum now becomes

$$\sum_t T(td'|n)\mu(t)/t = \sum_{t|(n/d')} \mu(t)/t,$$

where the fine print has now reverted to the bottom of the ad.

Next we want to relate this last sum to Euler's totient function, from the theory of numbers. The well known evaluation of the Euler function in terms of the prime factorization of an integer n is

$$\phi(n) = n \prod_{p|n}(1 - 1/p),$$

where the product is over prime divisors of n. If we multiply out all of the factors of the product we get an enormous alternating sum of reciprocals of various products of prime divisors of n. Those products run through precisely the square-free divisors of n, i.e., those in which no prime factor is repeated, and those are exactly the divisors of n on which the Möbius function is nonzero. What that all boils down to is that

$$\phi(n)/n = \sum_{d|n} \mu(d)/d.$$

If we substitute into the inner sum that we've been fussing over, and then substitute back in (2) we get the finalfinal result that exactly

$$F(n,k) = (1/n) \sum_{d|n} \phi(n/d)k^d \qquad (3)$$

different necklaces of n beads can be made out of beads of k colors.

There will be a brief pause while everybody lets $n = 4$ and $k = 2$ to see that the formula really gives $F(4, 2) = 6$...........OK, now that that's out of the way, let's try another simple case of the formula. This time, suppose n is a prime number. The virtue of that assumption is that primes have only two divisors, so there are only two terms in the sum for $F(n, k)$. After recalling that $\phi(n) = n - 1$ if (and only if) n is prime, we discover that

$$F(p, k) = (k^p - k + pk)/p.$$

Now the exact number is not the thing that is most interesting here. It's the fact that the right hand side is an integer. It has to be, because it's the number of ways of doing something! Thus, the numerator, $k^p - k + pk$ must be divisible by p, and therefore $k^p - k$ must be divisible by p, whatever the positive integer k might be. Well that is exactly Fermat's Little Theorem, and we got a proof of it as a spinoff from a combinatorial formula. The main point is that counting formulas must give integer answers, and if the answers don't *look* like integers then we may have discovered something interesting.

By the way, formula (3) must still be an integer even when n isn't prime, but it sure doesn't look like it. I wonder what that might mean...

§30. Excerpts from class, November 18 [notes by PMR]

> "No man but a blockhead ever wrote except for money."
> —Samuel Johnson, quoted in Boswell's *Life of Samuel Johnson*
> April 5th, 1776

Today we got an entirely different perspective on the whole ball of wax. Don began his fortnight's sabbatical by turning the stage over to one of Computer Science's most prolific authors: Professor Jeff Ullman. A large crowd had gathered to hear Jeff's advice on "How to get rich by writing books"—an illustration of one of the principles of cover design, he said: Attract people with something that isn't in the book at all.

Jeff started by talking a bit about the pragmatics of publishing—how the money flows. He kicked off with a back-of-an-envelope calculation. A book is a megabyte of text. Jeff can write perhaps two or three kilobytes of first draft per hour—say one kilobyte per hour of finished text. We can all train ourselves up to much the same performance, he asserted. So it takes around a thousand hours of labour to write a book. Now then, a typical CS text might sell for $40. A good book on a specialized topic, or a mediocre book on a general topic, might well sell 1500 copies in the US and 500 copies abroad. (These figures put the 200,000 copies of Don's *ACP* sold in the USSR into some perspective.) A 15% royalty is standard on domestic sales, a rather lower rate for foreign sales. All in all, our talented specialist or so-so generalist can expect to net maybe $8000 over his book's lifetime of perhaps five years. Of course, fame as well as fortune is to be gained through publication, but Jeff dismissed such non-financial motivations as being beyond the scope of his talk.

"I *told* you to be a lawyer. Or a doctor," someone's mother was heard to whisper. But Jeff forestalled a mass exodus to the GSB by going on to tell us how to make book-writing a going concern. Firstly, he said, it's quite feasible to double the royalty rate. CS authors have some leverage with publishers in that their books sell quite well—a publisher's costs are very sublinear in the number of copies sold, so he can afford to pay a lot more for a book that will sell 5000, instead of 2000, copies. What's more, a computer scientist often keeps his publisher's costs down by preparing his own camera-ready copies. Jeff is happy to tell you more about how to drive a hard bargain with your publisher—go and talk to him about it! He sees an upward trend setting in, with royalties exceeding 30%.

Secondly, you need to aim for ten thousand domestic sales; say two thousand a year for five years. That's 5–10% of the entire market in a topic like compilers or operating systems. There's nothing off-the-wall about this, provided you find the right niche: Let yours be the hardest book on the subject, or the easiest. Or the best. This wasn't so hard to do in the early days of CS, when there was a big demand for textbooks but only a few authors; it's certainly going to get harder as the field matures. If you're going for the big bucks, advised Jeff, choose a young and booming field—biogenetics perhaps.

Increase your royalties and sales, and your efforts can net you as much as a medium-grade hooker's: say $100 per hour. Top-notch computer scientists should aspire to no less.

A miscellany of tips

- Find a co-author or two. Co-authors won't save you any time, but they do help filter out your idiosyncrasies. Jeff said that when he writes alone "my own craziness takes over" and the book turns out a dud. He was just "too weird" in *Principles of Programming Systems*—although not too weird for the Japanese, who continue to buy it. His Database book went well, though probably because Chris Date's book provided a framework and the necessary "dose of reality." Filtering out oddball stuff has a big effect on quality. And since a textbook that is only marginally better than the competition will nevertheless grab the lion's share of sales, any small improvement is well worth having.

- Jeff never saw a book with too many examples. Use lots. Even a very simple example will get three-quarters of an idea across. A page or two later you can refine it with a complex example that illustrates all the "grubbies." But finding good examples— examples that illustrate all and only the points you are concerned with—is not easy; Jeff has no recipe. You must be prepared to spend a lot of time on it.

- Jeff endorsed Don's exhortation: "Put yourself in the reader's place!" If Mary-Claire concurs, we may even be convinced.

- Spend the day reading about a topic, and write it up in the evening. That way, you'll get the expository order right. You have an advantage over the experts because you can still remember what was hard to learn.

- Jeff often sees a definition in Chapter 2 and its use in Chapter 5. This just isn't the way readers work; it's essential to keep definitions and uses close together. Don't be ashamed to repeat yourself if that's what it takes.

- Those who can, do; those who can't, teach; those who can't teach, show off. Remember that the object of exposition is education, not showmanship.

- There is a tradeoff in using powerful mechanics to justify your methods; they may be too opaque. Jeff had to decide whether to spend 20 pages teaching asymptotic analysis in order to spend 5 pages applying its theorems, or whether just to say "It can be shown that ..." and refer his readers to another text. In the end he got around the dilemma by doing only the most basic calculations and proving nothing deep. In general, keep the level of your exposition down so that you can rely on your readers understanding it.

A couple of tactical remarks:

State the *types* of your variables. Talk about '...the set S ...', not about '... S ...'.

Jeff's English professor, now a leading poet, told him never to use the non-referential 'this'. Recognizing the dearth of poetry in CS, Jeff now forbids his students to use it either. 90% of the time it doesn't matter; the other 10% leaves your readers bewildered. One book presents four ideas in a row and then says "This leads us to consider ...". *What* leads us to consider?

Coping with the competition

Like it or not, book-writing is an increasingly competitive sport. But just because every other introductory Pascal text starts with '**write**' statements doesn't mean that yours has to starts with '**while**', just to be different. Don't slavishly imitate another's style, but don't avoid it either. Know the market, know thyself, and work out a compromise of your own. Don't hesitate to follow the crowd when they are all going in the right direction.

This last remark brought Jeff ("I am not a lawyer") Ullman round to the tricky subject of plagiarism. According to Prentice-Hall's *Guide to Authors*, imitation ceases to be the sincerest form of flattery and becomes something much more culpable if a reasonable person could not believe that you didn't have the other chap's book open in front of you as you wrote yours. That said, remember that you can't copyright ideas as such, but only ways of expressing them. Jeff shamelessly admits that his *Compilers* book borrowed another's table of contents and the general front-to-back expository scheme.

Jeff showed us a suspicious case in which an author had written "Knuth has shown ..." and then went on to quote more-or-less verbatim from *ACP*. The coincidence of notation is hardly conclusive, he said, but the identical use of italics is pretty damning.

Don here pointed out that his disciple had actually corrected a typo, for one sentence was in fact the exact logical negation of the other. But this book contained much worse examples of plagiarism: A dozen or so successive equations lifted straight from elsewhere. In these notes, names have been suppressed to protect the guilty.

Someone asked about second and subsequent editions. Jeff said that these will still consume a kilohour or so, although they'll go faster if you can use your earlier examples. But the financial advantages are very real: People stop buying a book when it has been out for five years, so publish a new edition and start the clock ticking again!

One person asked about writing survey papers—surely they will contain a lot of verbatim quotes? There's no problem since the writer is not presenting the work as his own, Jeff said. Besides, accusations of plagiarism hinge on financial loss, and no one writes technical papers to make money. But be explicit in your quotation if you feel more comfortable doing so.

Why don't expositions of CS make more use of analogy, asked someone, drawing an analogy with physics texts (which are planted thick with analogy, metaphor, and simile). Jeff thought it partly due to the nature of the subject, but encouraged us to use analogy where we are sure that the reader will get the point.

Asked about progress on *Parallel Computation*, Jeff confessed that it may never be finished: "That's another point about co-authors ...". Jeff left us, and Don, to reflect on his maxim:

"Never spend more than a year on anything."

§31. Excerpts from class, November 20 [notes by TLL]

Today's special guest lecturer was Leslie Lamport of DECSRC. Leslie, sporting a *Mama's Barbeque* T-shirt ("WALK IN — PIG OUT"), took the stage and gave us a very active lecture. (He clearly believes in one of his own maxims: "You've got to be excited about what you are doing.")

The first thing Leslie told us was that he would restrict his advice to the writing of papers (not books). "I have one thing to say about writing a paper for publication: Don't. The market is flooded. Why add to the detritus?" After the appropriate dramatic pause, he continued with, "But seriously folks, somebody has to write papers."

While we are asking ourselves if our own papers are worth writing, Leslie asks that we keep in mind two bad reasons for writing a paper:

The first bad reason is "to have a long publications list." Leslie says he would like to think that the people who are supposed to be impressed by a long publications list would be more impressed with quality than quantity. Admitting that this might not always be the case, he appealed to our own sense of integrity to police us where others' standards do not.

The second bad reason is "to have a paper published in a specific conference." Leslie has known people whose need to insert papers in specific proceedings is greater than their need to disseminate accurate information. This approach "sometimes leads to pretty sloppy papers." He told us that he knows of one case where the authors of a conference paper promised to send a correction, once they figured it out, to each conference participant.

Leslie recognizes *one* good reason to publish a paper: "You have done something that you are excited about."

Just how excited can you be and yet not publish a paper? Leslie was once told: "Judge an artist not by the quality of what is framed and hanging on the walls, but by the quality of what's in the wastebasket." Similarly, Leslie thinks that we should be judged on the "best thing that we have done that we decided not to publish."

Moving on to how we learn to write well, Leslie told us that learning to write is more like learning to play the piano than like learning to type. While both typing and piano-playing involve motor skills, a good pianist must spend much time studying music in its entirety; he must spend more time away from the piano than in front of it. Correspondingly, we should learn to write by reading. Leslie payed homage to Halmos and Knuth, but said that they can not match Fowles and Eliot: We should read great literature in order to learn how to write good mathematical literature.

We must know what we want to present before we can present it well. As Leslie said, "Bad writing comes from bad thinking, and bad thinking never produces good writing." We must keep in mind what we are writing—and to whom.

The question of audience is closely related to where a paper, once written, should be published. Appropriate places may be a Tech Report, a letter, a Journal, or the bottom drawer of your desk. (Don't really throw anything out: it is good to have the record, even if you don't publish your work.) How do we choose?

Journal articles should be polished and timeless. Conference papers can be a little rougher. Conference papers are appropriate for work that is "not yet ready for the archives." Technical reports (usually distributed by an institution) are good for work that is not even ready for the general world but still should be written up.

Leslie asks us to remember that "in each case, you still have readers. That tech report may some day turn into a Journal article. You've got to be excited about your writing."

As for the central theme of a paper, Leslie told us that he enjoys the Elizabethan use of the word "conceit" to denote a fanciful or cute idea around which a paper can be built. While such an idea can be a good catalyst as we begin to write, we should be willing to abandon it. After all, we use such metaphors or themes in order to present ideas—we should not allow them to intrude. The line between what can be called a conceit and the merely cute is a fine one. Beware of jokes. Just how funny will a joke be ten years after it was included in a Journal paper?

While jokes should be left out, examples are welcome additions to most papers. Leslie said, "It is better to have one solid example than to have a dry, abstract, academic paper." He also said that it is never a mistake to have too simple an example ("at least not for a lecture"). Demonstrating that "examples keep you honest," Leslie told us about a major revision of one of his published theories upon discovering that his original draft of the theory was not powerful enough to deal with the example that he wanted to use in his paper.

Expressing concern that people often "fix the sentence and not the idea," Leslie told us that we can be too concerned with details. For example, he tells us not to think about formatting when we are writing. ("Don't think about format. Do think about structure.") He suggests that whenever we have some detail, such as complex notation, we shouldn't write it out: We should use a macro.

Leslie discussed trends in notation, showing us a translation of Newton's *Principia Mathematica*. Newton stated his mathematical theorems in non-mathematical language that was very difficult to read. Instead of saying that something is inversely proportional to the square of the distance, we can get the point across better by saying that it is c/d^2.

Thus algebra has provided us with a tool for presenting the structure of a formula. But can't we improve present practice by making the structure of an entire discourse more clear? Leslie gave us a handout demonstrating two forms of a proof: a paragraph form and a form that looks like the tabular proofs that high-school students produce in Plane Geometry homework. (See §33 below.) Pointing out that the tabular proof is much easier to read, Leslie cautioned us that he was not talking about formatting, but the structure that the tabular form enforces. He says that writing proofs in such tabular "statement-reason" forms will help us clarify proofs that are to be presented in paragraph style. (The flip side of the handout also contains an example that Leslie did not have time to explain. The example shows some "bloated prose" that Leslie trimmed down by half.)

In discussing writing itself, Leslie said, "You should be excited about what you are writing and that excitement should show." Saying that this principle can especially be applied to first sentences ("You want something that leaps out at you"), he read us several first sentences from various compositions. The first sentence can be expected to be nontechnical and to represent an author's best effort. He was pleased with some of the first sentences from his own work and less pleased with others, but he was ecstatic about some of the first sentences he read us by T. S. Eliot or Allen Ginsburg. Thus it might be a good idea to ask ourselves: "What would T. S. have written, if he were writing this paper?"

What characterizes a good first sentence? Leslie says to "avoid passive wimpiness," but to be simple and direct. "Get right down to business." Of course, once you have hit your readers in the gut with your first sentence, you can't let them down with your second. Continuing in this vein, by induction, "When you come to sentence number 2079, you've got to keep socking it to them." (He illustrated this by reading an arresting sentence from the middle of *The Four Quartets* by T. S. Eliot, choosing the sentence at random.)

Leslie finished his lecture by saying, "I am not T. S. Eliot. I need to pay more attention to my writing. As do we all."

§32. How I changed my co-author's draft

In this section, we describe some of the highlights of the research area. We discuss some of the most significant, elegant, and useful algorithms, and some corresponding lower bound results. Since the literature in the area is vast and varied, we have found the selection and organization of these results to be a formidable task. We have chosen to simplify our task by restricting our attention to four major categories of results: shared memory algorithms, distributed consensus algorithms, distributed network algorithms and concurrency control. Each of these categories has a very rich research literature of its own, and we think that together, they provide a representative picture of work in the area. Still, our description is incomplete, since we neglect many other interesting topics.

In this section,

we discuss some of the most significant algorithms and lower bound results.

We restrict our attention to four major categories: shared memory algorithms, distributed consensus algorithms, distributed network algorithms and concurrency control.

Although we are neglecting many interesting topics, these four areas provide a representative picture of distributed computing.

§33. Toward structured proofs

Modified version of Corollary 3 from page 170 of *Calculus* by Michael Spivak:

Proposition If $f'(x) > 0$ for all x in an interval, then f is increasing on the interval.

PROOF: Let a and b be two points in the interval with $a < b$. We must prove that $f(a) < f(b)$.

By the Mean Value Theorem, there is some x in (a, b) with

$$f'(x) = \frac{f(b) - f(a)}{b - a}$$

But by hypothesis $f'(x) > 0$ for all x in (a, b), so

$$\frac{f(b) - f(a)}{b - a} > 0$$

Since $b - a > 0$ it follows that $f(b) > f(a)$. ∎

PROOF: Let a and b be two points in the interval with $a < b$. We must prove that $f(a) < f(b)$.

Statement	Reason
1. There exists x in (a,b) with $f'(x) = \frac{f(b) - f(a)}{b - a}$.	1. The Mean Value Theorem.
2. $f'(x) > 0$	2. By 1 and hypothesis.
3. $\frac{f(b) - f(a)}{b - a} > 0$	3. By 1 and 2.
4. $b - a > 0$	4. By choice of a and b.
5. $f(b) > f(a)$	5. By 3 and 4.

§34. Excerpts from class, November 23 [notes by PMR]

Nils Nilsson, latest in our line-up of megastar guest speakers, spoke on the subject of "Art and Writing." He began by showing us two photographs: Edward Weston's print of a snail-shell (strangely reminiscent of a human form), and Ansel Adams's "Aspens in New Mexico." Having thus set the artistic mood, Nils went on to talk about what this has to do with writing. Novels and plays are recognised as art; mathematical writing should also qualify, he said. Writing can be both art and communication; indeed, *real* communication happens only when writing is charged with artistic passion.

For Nils, a key word is *Composition*. Nils once took a course in photography from a teacher who declared that:

$$Composition = Organisation + Simplification.$$

This formulation made a lasting impression on Nils. It applies equally to writing as to photography. A quote from Edward Weston: "Composition is the strongest way of seeing." A typical artistic phrase, said Nils, but what does it mean? Some might say that Weston anticipated the findings of recent research in computer vision: The viewer must participate, construct models, form hypotheses. There are no spectator sports! Likewise, a photographer sees best when a scene is well-composed.

> "Life is very nice, but it lacks form.
> It is the aim of art to give it some."
>
> — Jean Anouilh.

But like all art, said Nils, writing should be fun. Just as the painter takes pleasure in the smell of his paints, so should the writer feel good when surrounded by the tools of his art: paper, ink, typewriter, word-processor, whatever. He must feel a thrill, as Don does when pin-pointing a reference. Another key word, then, is *Joy*.

But if writing is to be art, we must first master the craft. Only when our grasp of the minutiæ is perfect can we transcend technique and aspire to genius. Nils gave us a "broad brush" overview of some important points, along with some autobiographical tales.

1. **Start early.** Impressionable minds are best. Some people find that writing becomes a real compulsion; if this happens to you, then let the urge take over!

Way back in 1954 Nils took a Stanford course on "Scientific Writing." Writing an essay or two a week, he learned to become clear and organised—and got an A– for his paper on "Ionic Oscillations." Nils was pretty pleased, and thus began his career as a writer. In the Air Force, he discovered a growing urge to write a book about radar; he realises now that this was mainly a compulsion to get the material organised. By 1960 he had an outline of the book, but it never saw the light of day; after leaving the Air Force he joined SRI and got deeply involved with something else entirely (Neural Nets, as it happens).

2. **Write, rewrite, rewrite, rewrite … .** This dictum really is true, said Nils. It is the extremely rare artist who does not need to labour over and over on his work. Mozart was said to be an exception; his first draft was his final version. Beethoven,

on the other hand, rewrote his work over and over, and even then was never satisfied. As someone once remarked: "A work of art is never completed, only abandoned."

A member of the class quoted Robert Heinlein as saying that a writer must resist the urge to rewrite. (But then, Heinlein writes great thick books and pretty poor ones at that. Likewise, Barbara Cartland is said to wander about the house dictating her novels into a tape-recorder, whence they are transcribed and published. Of the literary qualities of her work, the less said the better.)

"Easy writing makes damned hard reading." Nils couldn't remember the source of this quote.* Hemingway put it rather more colourfully, which we blush to repeat here. Think of your early drafts as being like an artist's sketches, urged Nils: Be prepared to throw away nearly all of them. Neither are you done when the book is finished and on the shelves. Maurice Karnaugh (inventor of Karnaugh Maps) wrote to Nils after *Principles of Artificial Intelligence* was published and pointed out that the A^* algorithm as Nils had defined it would fail on a certain graph. This led to a correction in a subsequent edition.

Never let anything you write be published without having had others critique it. A university is a good environment in which to get feedback on your work, though you may need to give some thought to the timing of your requests for comments (unless you have infinite resources of willing readers). Nils told us about the time that he thought he had a neat result in non-monotonic reasoning and circumscription. He wrote it up and sent it to John McCarthy, who passed it on to Vladimir Lifschitz, who discovered that Nils's derivation "appeared to contain an oversight ...".

Nils always tries to teach a course on a topic at the time that he is writing it up—it's ridiculous to inflict your ramblings on the world unless you are prepared to do this, he said.

Nils decided that since he had the whole book on-line, he would take a crack at publishing it himself. He and his wife Karen set up the Tioga Publishing Company. One big advantage about this cottage-industry approach is the ease with which the author can make changes in subsequent editions. Karen went on to become a full-time publisher; Tioga's theme has now changed from AI to nature and the environment. So Nils considers himself pretty well "vertically integrated" in the world of books.

3. **Read.**   Read a great deal; it'll sharpen your style and get your critical faculties working.

4. **Model the Reader.**   Déjà vu. This should be obvious, said Nils, but there's really a lot to it. Ask yourself what the reader's primitives are, and write with them in mind. In fact, the whole issue is so complex and important that Nils likes to operationalise it with AI-type "dæmons." Any number of these have to be running in the background as you write, catching errors and providing constructive criticism. You have to be asking all the time; "How is the reader going to misunderstand me here?" You must

---

* "Easy reading is damned hard writing." — Nathaniel Hawthorne. ("Just lucky to find it." — DEK.)

automatically insert forests of guidelines to keep him on track. You develop these dæmons by practice—it's a kind of motor skill, like playing tennis or riding a bicycle. A split infinitive should really jar, Nils said: "It's got to light up in red!" The dæmons have to run automatically; you can't be consciously checking a list of rules all the time. Besides, if writing is to be fun, it can't be compulsive!

5. **Master the Medium.** You need a good vocabulary, though this needn't mean a huge list of big words. There are issues other than pure language: indexes, tables, graphs, and how to use them to best effect. As Don pointed out earlier, we can use typography to make important distinctions, as with the typewriter font for logical formulæ.

In the future, said Nils, it's clear that reading and writing will be far more interactive processes—*The Media Lab* is not all hype. It's not clear yet what will prove necessary or useful; just as it took several centuries to invent the index, it will probably take us a long time to identify the "stable points" offered by our new technology. We in the audience are at the cutting edge of these experiments.

6. **Master the Material.** There's a lot of internal feedback involved in writing; one comes to understand the material in a new way on trying to organise it for publication. Nils drew this diagram :

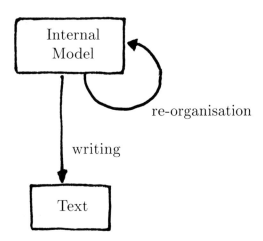

As Mary-Claire said on Wednesday, "How do I know what I mean until I hear what I say?" Even Nils sometimes finds himself thinking "I don't believe that!" when he hears himself lecture. I am reminded of the (true) story of a professor who was always seen to take a pad of blank paper with him when he delivered a talk. When asked what was for, he replied: "Why, if I say anything good I'll want to write it down!" So go to lectures and classes, give talks. All these things help modify your internal model and get things into shape.

> "In a very real sense, the writer writes in order to teach himself, to understand himself; the publishing of his ideas, though it brings gratifications, is a curious anticlimax."
>
> — Alfred Kazin

7. **Simplify.**   Lie, if it helps. You can add the correct details later on, but it is essential to present the reader with something straightforward to start off with. So don't be afraid to bend the facts initially where this leads to a useful simplification and then pay back the debt to truth later by gradual elaborations.

"Another noteworthy characteristic of this manual is that it doesn't always tell the truth."

— Don Knuth, *The TEXbook* (page vii)

"Everything should always be made as simple as possible, but not simpler."

— Albert Einstein

Ted Shortliffe did a great job with Mycin, Nils said. But with 20/20 hindsight he might have done better to invent a simplified system for expository purposes. For example, he could have demonstrated the backward-chaining techniques and only later dealt with "certainty factors."

By using simple examples we can get ourselves on the winning side of the 80-20 rule: we can convey 80% of the truth with only 20% of the difficulty. Mathematicians, of course, like to go the other way: They never state a theorem in three dimensions if it can be generalised to n. Such terse elegance can be painful for the reader.

8. **Avoid Recycling.**   With on-line text and sophisticated editors (I refer to software, not the mandarins behind *Scientific American*) it is very tempting to re-use portions of old material. Resist the temptation. Almost certainly you are writing in a new context, with a new emphasis. Hopefully you are older and wiser, and perhaps even a better writer than you were when the old material was written. So do rewrite it, it's worth the extra effort.

9. **Aim for Excellence.**   You've got to keep shooting for perfection, even if you'll never get there. What the Great have said on this:

"We are all apprentices in a craft where no-one ever becomes a master."

— Ernest Hemingway

"Someday I'll build the perfect birch-bark canoe."

— John McPhee

"Someday I'll write the perfect AI textbook."

— Nils Nilsson

"Ah, but a man's reach should exceed his grasp, or what's a heaven for?"

— Robert Browning

"The message of these books is that, here in the 80s, 'good' is no longer good enough. In today's business environment, 'good' is a word we use to describe an employee whom we are about to transfer to a urinal-storage facility in the Aleutian Islands. What we want, in our 80s business executive, is somebody who demands the best in everything; somebody who is never satisfied; somebody who, if he had been in charge of decorating the Sistine Chapel, would have said: "That is a good fresco, Michelangelo, but I want a better fresco, and I want it by tomorrow morning."

— Dave Barry

§35. Excerpts from class, November 25 [notes by TLL]

Don opened class by introducing guest lecturer Mary-Claire van Leunen and by giving us the title of her talk: "Calisthenics." Mary-Claire opened her talk by telling us a story.

Many years ago Mary-Claire was a frequent passenger on the Chicago bus system. The neighborhood where she boarded her #5 bus was a gathering spot for "bummy guys." All of these guys were interested in money: Some begged, others peddled. Among the peddlers—hawking wares ranging from trenchcoats full of watches to freedom from the peddler's presence—was a man whom Mary-Claire patronized quite regularly. He sold pencil stubs (obviously collected from trash bins); but Mary-Claire said his patter was charming enough to rate one or two purchases a week.

"These pencils are magic pencils," he would say. "Buy a magic pencil. Only 25 cents."

"What's a magic pencil?" would come the expected response.

"With this pencil, you can write *the truth*."

Inevitably, someone would pipe up, "But I can write lies with it."

"Oh, you can break the magic. But if you really believe, you can write *the truth*."

Mary-Claire sees this as the wonder and the motivation behind the craft of writing: If you work hard, you can explain a new truth to someone you will never meet—perhaps to someone who will live after you are dead.

Such a vocation requires preparation. The Composition Exercises that Mary-Claire has given us (see §36 below) were designed to help us become as strong as we can. Our readers are more likely to be tolerant of a few weaknesses if they are surrounded and supported by strength.

Mary-Claire has given these exercises to students before, but preparing this draft for our class pushed her to really *write* the exercises. The copy that she referred to over the TV monitors was slightly different than the copies that we have been given. Mary-Claire, hoping that these differences represent improvements, invites us to suggest further improvements to the draft. (She says that she might publish something that evolves from this draft—but probably not soon: She is not a fast writer.)

The first set of exercises, labeled "Vocabulary," is designed to increase our command of just that.

The first of the pair is an exercise that was done by little Greek boys: Taking a composition and swapping all the old words (nouns, verbs, adjectives, and adverbs) for new ones. What is the effect of these changes? What happens when a vulgarism is used? When a hoity-toity word is used?

The second vocabulary exercise, the writing of a thesaurus entry, is best done over a week. After several days of slowly adding to our set of synonyms, we should compare our entry with an entry in our thesaurus. (Everyone needs at least one thesaurus and a good unabridged dictionary. In addition to more than one kind of dictionary, Mary-Claire recommends Sidney Landau's book *Dictionaries* to help us understand how to best use our dictionaries.)

"Syntax," the next set of exercises, deals with syntactic mastery. Mary-Claire says even though vocabulary improvement is more often considered than increasing our command of syntax, syntactic armory improvement is more important. She says that most of the time we will use our basic three to four thousand words; we must use them in the most interesting way possible.

Speaking of using words in interesting ways, Mary-Claire has been reading the first draft of our term papers. There must be room for some improvement there: Her first comment was, "Nobody sits down to write a boring paper." How can we tell when something we write is "syntactically impoverished"? She gathered some statistics that might help us get the right idea.

One of her tricks was to study the first 10 complete sentences on the third page of every paper. First she charted the average length of the 10 sentences: They varied from 15.6 words per sentence to 24.4 words per sentence. Mary-Claire says that any of us with averages under 20 words per sentence are in the correct range for adult writing. (But the writer with the 24.4 average had better have results pretty wonderful, to compensate for the extra work that it takes to read his paper.)

Sheer variation in sentence length is one indication of syntactic variation and appropriate pacing. With 10 sentences we should be aiming for 9 or 10 different lengths. The samples from our papers yielded 6 to 9 different lengths. The difference between the word count on the shortest sentence to the word count on the longest varied between 17 words to 37 words. The ideal chart of sentence lengths should look like a bell-curve centered around 15 to 18 words per sentence.

She asks us to note that we did not have enough short ("and punchy") sentences. A few long sentences are also important. She said, "A well constructed 46-word sentence is not a difficult beast, but it had better not be the your crucial point." We should remember that we have a responsibility to emphasize and deemphasize our points to the reader; long sentences are one method of deemphasizing a point.

Beyond the word counts, she looked at the the templates used to construct our sentences. For example, she found two writers who would appear to be similar if we just looked at their sentence length average and variation, but who had quite different methods of constructing their sentences. One of these writers used the same sentence construction for almost every sentence (adverbial + subject + transitive-verb + object), and the other used many different styles of construction. But the second writer was not free of flaws. He had two sentences in a row with a full independent clause followed by a full parenthetical independent clause. Mary-Claire says that we must learn what syntactic usage is unusual so that we do not overuse it.

One syntactic trick not normally thought unusual was startlingly absent from our papers: tight parallels. The use of two adjacent sentences with exactly the same construction is an effective way to communicate similarity to our readers. Mary-Claire is curious how we could all avoid this technique. Perhaps it is an artifact of the way that students of our generation were taught?

Given this motivation to "increase our syntactic muscle," Mary-Claire led us back to discussing the "Syntax" exercises. She passed over the first two exercises as obvious, but a few comments were made on "periodic" sentences.

We who have had mathematical training might be tempted to guess that a periodic sentence is one that repeats cyclically, but we would be wrong. Periodic sentences are those whose grammatical and physical ends coincide: We must get adverbials out of the final position. For example, a verb that is intransitive must end the sentence. Period.

Periodic sentences are not really appropriate in our kind of writing; they are a high literary form. Even though such a sentence form is more frequently encountered in church than in conference papers, the use of periodic sentences will heighten our awareness that we can control sentence structure.

The next exercise has us recast a sentence so as to change emphasis. What are the emphatic positions? The front of the sentence and the back. ("The middle of a sentence is sort of a slum.") But she says not to take her word for it; we should write sentences with varying emphasis and find out for ourselves.

Mary-Claire says that the last syntactic exercise is "incredibly wonderful": Write nonsense. Write a completely unrelated stream of thoughts with the correct glue: Words like "thus," "therefore," and "as we can see." She says this is a fun exercise to do after a couple of drinks. (Maybe we need a class lab?)

Moving on to exercises labeled "Manual labor," Mary-Claire told us that these should logically come first, but she wanted to woo us with the logical stuff before we ran into the weird stuff. Why is it important to use different methods to copy other people's writing? Because writing—and even reading—is partly a manual process. Mary-Claire typed out large sections of our papers as she was analyzing them. She said that if you tie a baby's hands behind his back, but give him otherwise adequate mental stimulation, he will not learn to speak well.

When we want to read a passage of text seriously, such manual labor can help us slow our brains down until we can give the passage the consideration it deserves. (W. H. Auden said that the proper way to show contempt for a poem is to copy it on a typewriter; the proper way to show admiration is to copy it in longhand.) Memorization and recitation can also help us to be able to read *word* by *word*. ("Make yourself into a book that you can take to prison should worse come to worst.")

Mary-Claire is aware that we may not buy this "manual labor" technique at first, but she asks us to take it on faith. She took the technique on faith for ten years and then wrote a sincere Thank-You to the person who told her about it. Ten years from now we can write her and tell us our opinions.

Most writers are aware how important the manual part of composition is: They have very rigid restrictions on how they compose. ("Oh, I can only write on yellow pads with a fountain pen." Mary-Claire says we should be able to compose on a cocktail napkin.

While discussing the section labeled "Frozen sounds," Mary-Claire told us about reading aloud to her students their own writing. Some students were chagrined; others glowed. She says we should form partnerships with other novice writers: Read and listen to each other. But she cautions that a little goes a long way. If the writing is good, we can live on that joy for quite a while; if the writing is bad, we won't be able to stand it for very long.

At this point in the lecture, Mary-Claire noticed that very few minutes remained. So her comments on the final exercises were limited to those that she thought were the most important.

Concerning the "Marks on paper" exercises, Mary-Claire quoted from E. M. Forster: "How do I know what I mean till I see what I say?" We need to remember that writing is "the most forgiving medium known to man." We can work on it until we get it right.

Rushing past the "Stance, voice, and tone" section, she told us that she borrowed techniques from speech therapists—who ask patients to exaggerate their defects until they understand just exactly what characterizes their defects. For example, she says, "If any one has ever told you that you are 'breezy,' write something truly off the wall."

She told us that the sections labeled "Observation," "Same as and different from," and "Invention" are less important for us than for pure writers. Our discipline provides the glue that writers with more freedom have to manufacture from scratch.

She warns us that the "Scansion" exercises are hard, but very important. She realizes that she may have trouble convincing us that we need to write verse in order to learn to write mathematics, but once again she says, "Trust me."

She reminded us that the "Precis" exercises were touched upon by Leslie Lamport in his talk. At some point we cannot reduce the word count of a piece of prose without changing the structure of that prose. (We should never change the meaning, but we will have to dispense with some details.) This point comes at different percentages of decrease— depending on the flabbiness of the original text.

The final exercises she discussed, "Nearly real," are aptly named. They really are very much like real writing. For instance, Mary-Claire says that "Writing a joke is exposition at its purest. Things aren't funny unless they are well written."

She suggests that we try "Ben Franklin's exercise," rewriting a passage of someone else's from memory and limited written hints—but that we try it with Don's writing. When we have finished, what do we like better about Don's version? What do we like better about our own?

Before the cameraman could shoo us out of the room, Mary-Claire reminded us once again that these exercises are "very hard work." She closed with, "I hope they will serve you as well as they have served me."

Some of us surely hope the same.

§36.

Composition Exercises

*** Draft ***

Mary-Claire van Leunen

Unless you plan to do nothing else but composition exercises, there are enough here to last you for the next decade. I had a wonderful student who did nearly all of them in a year, but he really did do nothing else. Some of the exercises are quite deep, and you might easily be able to do them again and again in different guises for the rest of your life. I've done all of them myself.

Many of these exercises tell you to take a passage of such-and-such a length and work some kind of transformation on it. Whose passage should it be, your own or someone else's? Either; or rather, both. You can learn different things by doing the exercise different ways. If a piece of your own writing is still fresh enough in your mind so that you can remember what problems you were trying to solve as you wrote it, the experience of working completely arbitrary changes on it can be exhilarating, not unlike setting dollar bills on fire.

Vocabulary:

1. Replace.

    Take a passage five pages long and replace at least three words in every sentence with others that mean approximately the same thing. (= "Get into your hands a 1500-word portion of a written work and swap out of every sentence a minimum of three words, substituting others without changing what's being said." = "Select a longish section from something you've been reading and change the vocabulary of every sentence without changing the signification.")

2. Multiply.

    Choose a word and write a thesaurus entry for it -- all the words at every level of diction that mean approximately the same thing. Compare your entry to the entries in which the word actually appears in some real thesaurus.

Most writers like to have several good desk dictionaries and at least one good thesaurus. In addition, you might like a book by Sidney Landau called *Dictionaries*. It has helped me understand how dictionaries and thesauruses get made and thus how to use them better.

Syntax:

1. Transform.

    Take a passage of five pages and transform every sentence so that it says approximately the same thing in different syntax. Change the vocabulary as little as possible. (= "Taking a five-page passage, transform every sentence to say an approximation of the same thing in different syntax." = "Can you transform every sentence in a passage of five pages so that approximately the same thing gets said in different syntax?")

2. Build tight parallels.

    Write a sentence containing a tight parallel: a pair of structures that match perfectly in the number and kind of all their parts and subordinate structures. Push yourself till you can construct sentences that contain tight parallels in which each member is fifteen or twenty words long.

3. Be periodic.

    Rewrite a non-periodic sentence or pair of sentences as a periodic sentence.
    (A periodic sentence is one whose grammatical and physical ends coincide -- one with no adverbials to the right of the predicate. The first sentence in this paragraph is non-periodic; a periodic version might read: "Rewrite as a periodic sentence a non-periodic sentence or pair of sentences.") Write an entire paragraph in periodic sentences; an entire page.

4. Emphasize.

    Recast a sentence so as to express the same meaning but emphasize a different point.

5. Write nonsense.

    Write a paragraph of coherent, tightly structured nonsense --- all the connectives and labels in place, but no meaning. (Made-up words not allowed.)

Manual labor:

1. Copy I.

    Copy out a passage of your own writing or someone else's with a pen; with a pencil; with a crayon; first with your left hand and then with your right; with a manual typewriter; with a word-processor.

2. Copy II.

    Copy out a passage from something you like; from something you dislike; from something you find difficult to read; from something you find laughably easy; from someone you'd like to imitate; from someone completely unlike you; from something written a hundred years ago; from something written last year; from something scrawled off in haste; from something overwritten and finicky.

The first of these copying exercises is exploratory and interesting, and I certainly recommend that you do it, but the second is in another class altogether. Copying as a means of close reading is an inexhaustible source of information. Word-processing has temporarily confused writers about the connection between their hands and their brains. What you do with your hands is the easy part. Use it to support the hard part. The hard part which is what you do with your brain.

Frozen sounds:

1. Transcribe.

    Make a tape recording of five minutes of radio news and transcribe it. Transcribe five minutes of a public lecture; five minutes of dialog from a television show; five minutes of ordinary conversation among three or four people. Talk extemporaneously into a tape recorder for five minutes and transcribe that.

2. Listen.

    Read aloud a page of your own writing. Ask a friend of yours to read it aloud. Ask a second friend. Ask a stranger.

Learn to mutter aloud what you're writing as you write it. It's only a minor eccentricity, and there's no more efficient way of checking for both cadence and tone.

Marks on paper:

1. Close your eyes.

    Compose a paragraph with your eyes shut. Start over again from the beginning on a fresh piece of paper as often as you like, but don't peek.

2. Tabulate.

    Write a sentence with two conditionals ("If it rains and if Peter arrives on time ..."); rewrite it as a table. Write a paragraph with several linked conditionals; rewrite it as a table.

3. Caption.

    Describe a picture in a single line that fits under the picture exactly; in two lines that fit under the picture exactly.

Stance, voice, and tone:

1. Change stance.

    Rewrite a textbook explanation as a personal letter to an intelligent child -- a beloved niece or nephew, for instance. Write a description of this morning's events as a letter to your spouse or lover; rewrite the description as a letter to an old high-school teacher of yours; rewrite it yet again as a report to an examining psychiatrist; to an anthropologist; to a police inspector; to a reporter from People magazine.

2. Take both sides.

    Write a vigorous, closely reasoned argument for some small household economy like re-using plastic bags or turning mattresses; now write a vigorous, closely reasoned argument against.

3. Hyperbolize.

    Describe your current dwelling as an unscrupulous relator would; describe your most recent meal in restaurant-menu prose; describe an object on your desk as if it were for sale by mail order.

4. Euphemize.

    Describe the symptoms of severe gastroenteritis accurately but without recourse to vivid language. Describe human sexual intercourse in the diction of a knowledgeable prude. Describe an employee's forced resignation for incompetence in language that attempts to leave no opening for a libel suit.

5. Obfuscate.

    Take a passage of simple prose and rewrite it so that the same ideas seem obscure and difficult.

6. Pontificate.

    Take a straightforward passage written in the first person and rewrite it so as to make the author seem pompous and self-important.

7. Vacillate.

    Take an argumentative passage and inject it with doubts, quibbles, and hesitations.

8. Strengthen; vitiate.

    Find a weak, flabby paragraph and rewrite it, inventing ideas and details where necessary, to make it vigorous and strong. Now do the reverse: Find a strong paragraph and weaken it.

9. Change tone.

    Look through a magazine or a newspaper for a sarcastic letter to the editor; rewrite it to make the same point but without the sarcasm. Find a short factual piece; rewrite it as the preamble to a petition asking for some action on the facts.

Observation:

1. Expand and contract.

    Write a paragraph describing some small event of the last day -- fixing your breakfast, or catching the bus, or buying a newspaper. Expand the description to five pages. Now cut it back to a paragraph again and compare the new paragraph to the old one.

2. Rethink.

    Describe a favorite food by its appearance alone; describe only the sounds in the opening credits for a movie; categorize and describe the objects on your desk by texture; by color.

3. Sensualize.

    Choose an object and describe it by sight; by sound; by smell; by taste; by touch. Choose an event from your daily life and describe it as a sensory experience.

4. Louis Agassiz's exercise.

    Put a green leaf or a flower on a plate and describe it every day for two weeks. (The original version used a fish and took two months.)

Same as and different from:

1. Compare.

    Choose two unlike things and build a simile capturing some point of similarity between them. Choose two similar things and explain how they differ.

2. Analogize.

    Invent an extended analogy that would help an illiterate understand what a library is good for; that would help a child understand getting fired from a job; that would help a city-dweller understand the agricultural year.

3. Differentiate.

    Choose a ten- or fifteen-word entry in a thesaurus and explain how the words differ from one another.

Invention:

1. Combine words.

    Choose two words at random from a dictionary and write a sentence that uses both of them; choose three words at random and do the same.

2. Categorize.

    Take twenty nouns at random from a dictionary and arrange them in categories. Write an explanation of your scheme for arranging them.

Usually we don't need to do pure invention; we start from something, even if it's only "What I Did on My Summer Vacation." Much of learning a discipline is learning how to do invention -- how to recognize the kinds of ideas that make that discipline go forward and how to get yourself into position to have such ideas yourself.

Another part of learning a discipline is learning what you don't have to invent because it's already been done for you. The forms for taking advantage of that backlog of ideas vary from one discipline to another, but the underlying habits of thought are similar. The best set of exercises I have ever seen on those habits is at the end of the section called "External Aids to Invention" in Edward P. J. Corbett's *Classical Rhetoric for the Modern Student*.

Scansion:

1. Versify.

    Render a newspaper story in couplets of iambic tetrameter; in triplets of dactylic hexameter. Render a recipe into rhymed vers libre. Render an expository passage as a ballad. Render a short argumentative passage as a villanelle.

2. Sonnetize.

    Write a new sonnet every day for a week. (Be sure to throw these sonnets away.)

3. Explode.

    Take a piece of metric verse and expand every line by one foot without altering the meaning -- from tetrameter to pentameter, for instance, or from pentameter to hexameter.

Verse-writing is to other composition exercises what lifting hundred-pound weights is to touching your toes. But you must honor conventional rhyme and stress in order to get the benefit of writing verse; otherwise you'll cheat yourself by writing near misses. For help on words like "villanelle" and "hexameter," get a prosody handbook; get John Hollander's, and you'll find yourself reading it for fun.

Precis:

1. Reduce.

    Choose a passage and count the number of words in it. Reduce them by 5% without changing the meaning; by a quarter; by half.

2. Abstract.

    Describe in no more than ten sentences the content of an article; of a book.

Nearly real:

1. Flip.

    Take a paragraph and rewrite it so that the last sentence comes first and the first sentence comes last. The middle will have to be completely rewritten, but try to change the first and last sentences as little as possible.

2. Replace.

    Rewrite an ordinary paragraph so as to enforce a leisurely, ruminative pace on the reader; now rewrite it the other way, to make it seem unusually quick and light and tripping.

3. Crunch.

    Write a 300-word description of some concrete physical object without using any adjectives or adverbs; write a thousand-word description without any.

4. Unpack.

    Take a metaphorical passage in either verse or prose and rewrite it as a series of flat-footed prose comparisons.

5. Define.

    Write dictionary definitions for a common word like "hand" or "mean" or "find." Compare your definitions to those in several dictionaries.

6. Exemplify.

    Choose an abstract noun like "generosity" or "fortitude" and describe three or more instances of it. Impose an order on the instances and explain the order.

7. Explain.

    Rewrite a simple sweater pattern to meet the needs of someone who has never knitted. Rewrite a chocolate-cake recipe for someone who has never cooked. Rewrite directions on how to set ignition points for someone who has never driven a car.

8. Instruct.

    Write a set of instructions on how to draw some fairly complicated object without even naming it or any of its parts -- a Christmas tree with ornaments and a star on top, for instance, or a house with a chimney, windows, doors, and foundation plantings.

    Try your instructions out on a friend.

9. Write a joke.

    So go ahead, write a joke.

10. Translate.

    Buy a book in a language you don't know and a bilingual dictionary for the language. Translate passages.

11. Ben Franklin's exercise.

    Take a passage of someone else's, three or four pages long, and reduce it to a set of one- and two-word hints to yourself about the contents, each written on a separate piece of paper. Jumble the hints, put them in a box, and take them out again after three weeks. Arrange them and reconstruct the passage. Compare your reconstruction to the original.

12. Push.

    Write sentences at a deliberate pace for five minutes without repeating yourself, without writing nonsense, and without stopping. Increase the time gradually till you can do this exercise for twenty minutes.

Tropes:

In addition to doing all these exercises, my student and I also worked our way through Richard Lanham's *Handlist of Rhetorical Terms*, writing an example for every rhetorical figure listed. Contrary to what I had expected, writing an example for every figure in Lanham turned out to be quite shallow. I believe that merely thinking about our doing it will give you every bit as much benefit as doing it yourself.

23 November 1987

Books Mentioned

Benjamin Franklin.
The Autobiography.

Edward P. J. Corbett.
Classical Rhetoric for the Modern Student.
Oxford University Press, second edition 1971.

John Hollander.
Rhyme's Reason: A Guide to English Verse.
Yale University Press, 1981.

Sidney I. Landau.
Dictionaries: The Art and Craft of Lexicography.
Charles Scribner's Sons, 1984.

Richard A. Lanham.
A Handlist of Rhetorical Terms: A Guide for Students of English Literature.
University of California Press, 1969.

§37. **Excerpts from class, November 30**   [notes by PMR]

> *During the whole of a dull, dark, and soundless day in the autumn of the year, when the clouds hung oppressively low in the heavens, I had been passing alone, on horseback, through a singularly dreary tract of country; and at length found myself, as the shades of evening drew on, within view of the melancholy Terman Engineering building.*
>
> —E. A. Poe (amended)

Don, like Mary-Claire, scans the pages of *The New Yorker* for choice malapropisms to entertain us. In its columns the law firm of Choate, Hall, & Stewart had been rendered as Choate, Hall, Ampersand, and Stewart, presumably by a journalist receiving dictation over the telephone.

We also saw a splendid dangling participle from the same source:

> "Flavor and texture of cooked okra are different from other vegetables. We usually don't eat it raw, but in judging at fairs, I frequently taste a slice of a pod to check maturity and condition. In soups, it is used as a thickening agent. When fried, I love okra."
>
> [When sober, can't stand the stuff. —*The New Yorker*]

Don announced that he had good news and bad news for us. He gave us the good news first. Mary-Claire is to speak again on Wednesday. Also, Don finally got up the courage to ask Paul Halmos to appear in our guest spot; he readily agreed and will speak next Wednesday (9th December). This talk should be a fitting climax to the course. And a week from today (Monday, 7th December) we will hear from Rosalie Stemer, a copy-editor for *The San Francisco Chronicle*.

Having thus softened us up with these cheerful tidings, Don delivered the Bad News: The first drafts of the term papers were, well ... "their content was not one hundred percent pleasing to your instructor." What makes a professor's life worthwhile? The knowledge that he has succeeded in teaching something. In particular, there's a joy in the thought that a student was able to do something that he couldn't have managed without the professor's help. Don confessed that this joy did not run through him as he read our drafts. Indeed, he could almost think that many of them were written before the class started. Have we been relaxing too much, he wondered? Has our writing in fact changed at all? Have we learnt nothing? Disturbing thoughts, he said.

Of the thirteen papers submitted, eleven were sprinkled with wicked whiches—at least two in each. Don himself has been guilty of these in his time, and of course there is no-one like a convert for rooting out heresy. But these are the 80s and we are supposed to be sensitised to these things. And heaven knows, we've talked about this issue enough in class! So what is he to think about this landslide of carelessness? Shaking his head, Don declared that we left him with no alternative—he would have to resort to the ultimate sanction: a quiz.

In keeping with Honor Code protocols, Don left the room while we each wrote a sentence that used a 'that' correctly where a 'which' would have been wrong, and another complementary sentence—which used a 'which' correctly where a 'that' would have been incorrect. A minute passed. And then another minute. There was little to hear but the scratching of pencils and the beating of hearts.

Don returned. We spent a few minutes looking at the various examples that the class had come up with, some correct and some incorrect. By and large, the class redeemed itself by the creative solutions that were submitted:

> All the students that know when to use 'which' and 'that' will pass the quiz. The exam, which took place at the beginning of class, was not difficult.

> A paper that uses two whiches improperly does not demonstrate that the author hasn't learned anything. My first draft, which was written this summer, had a million of them.

> Beware of examples that are misleading. My term paper, which contains many wicked whiches, is otherwise not too bad.

CS-types just love self-reference, it seems.

Is it not true, TLL asked of Mary-Claire, that people invariably get their whiches and thats right when they speak? Mary-Claire replied that people almost never say 'which' improperly in general speech—it's only when they feel under pressure that they resort to this unnatural diction. So unnecessary use of 'which' really conveys a bad tone in your writing; it makes you sound nervous. (Conversely, on paper we can often fool our audience into thinking that we are a lot more comfortable than we really are).

Don observed that all translations of the Bible are strewn with erroneous whiches. ("Thou shalt not suffer a wicked which to live," he might have said.) A clamour of voices pointed out that Fowler is quite clear about the rule. True, but it was never enforced until the late 70s, Don countered. It seems particularly strange, he said, that the *New English Bible* should commit this error, as its editors take great pride in the literary qualities of their text. Mary-Claire resolved this mystery: Apparently our "oldest and closest allies" on that far-off island regard this whole issue as unmitigated nonsense!

Don made a final plea to us: "You all keep your text online, so it's very *very* easy to locate all your whiches and check them. Please don't cause your instructor any more pain on this score!"

Sneaky Don had saved one more item of good news to lighten our spirits after this depressing interlude: A letter from Leonard Gillman, editor of the Seirpiński proof over which we had laboured many moons ago. Professor Gillman was lavish [not 'fulsome', according to page 23 of his book] in his praise of our suggestions, and is now working on an improved write up. Particular credit went to Student B, of course. Gillman is an Emeritus Professor as of three months ago—Don drools to think of all the free time he must have.

We moved on. Don claimed to have discovered a new (?) rule only by seeing it broken in three of the papers he read. It is this: The text should make sense if we read through it omitting the titles of subsections. So, for example, don't say:

> **2. Contour Integration.** This technique, invented by Cauchy, is used ...

Rather, say:

> **2. Contour Integration.** The technique of integrating along curves in the complex plane, invented by Cauchy, is used ...

The point is that the subheading should not be referred to explicitly or regarded as an "integral" part of the text. Think of it as some kind of marginal note or meta-level comment.

We spent the rest of the time looking at the draft of a paper about graph theory written by Ramsey Haddad and Alex Schäffer. Firstly, Don pointed to their careful attention to definitions. This is particularly important in graph theory as various authors use terms inconsistently. A *path* may or may not be the same thing as a *simple path*. At least one writer uses the term *walk* to make a distinction here. So it is necessary to define one's graph theory terms clearly right at the start, even the most basic ones. Remarkably, there was a time when the symbol '=' was not in general use. Fermat never used it, preferring always to write 'æq' or 'adæq' or fuller Latin words like 'adæquibantur' that these terms abbreviate. So in those days you would have to define the symbol '=' at the beginning of your article if you intended to use it. (The equals sign was invented by Robert Recorde in his *Whetstone of Witte*, 1557, but it did not come into general use until more than a hundred years later. Descartes used '=' to mean something completely different.) The moral: Ask yourself what background your readers share, and what they may or may not have in common. "Be aware of what's diverse in your readership".

We saw a somewhat intimidating multi-part definition. It would become less formidable to the reader if shortened. In this particular case, the expression

$$W_1 \to W_2 \xrightarrow{*} W_3$$

could have been condensed to

$$W_1 \xrightarrow{+} W_3$$

since W_2 is used nowhere else. (In the Haddad-Schäffer paper, '*' means 'zero or more' and '+' means 'one or more', but let's not worry about that here.) Try to be succinct, said Don: "Less is more."

It is important to be consistent in your use of terms, and you need to be especially careful about this when working with co-authors. In this paper, one writer talked about 'dominators' and the other about 'parents', referring to the same concept. (Freudian slip?) A related issue: Don't define terms that you never use. Don recalled Feynman's complaint about New Maths: you are taught the symbols ∩ and ∪ in second grade, but you don't use them in any nontrivial way for seven years.

Next came a tricky question of tenses. "Gabow and Tarjan[Gab83] show that for many algorithms that had such a multiplicative factor in their worst-case complexities, the multiplicative term can be removed." Here 'had' should be 'have'; an algorithm lives forever, and its worst-case complexity is a timeless fact about it. However, the *problem* solved by an algorithm can have different known complexities at different times; therefore 'had' would be okay if 'algorithms' were 'problems'. (The quoted sentence also exhibits other anomalies. A 'multiplicative factor' is not also a 'multiplicative term'; factors are multiplied, terms are added. Also the logic of the sentence can be unwound to make the point clearer: "Gabow and Tarjan have shown how to improve the algorithms by removing such a multiplicative factor from the worst-case complexities in many cases [Gab83].")

We talked about abbreviations for bibliographic references. Don didn't like the lack of space before the bracket in "...Tarjan[Gab83]..."; neither does he like this kind of thing: "In [Smith 80] it was shown ...". References should ideally be parenthetical; we should be able to read the sentence ignoring them and still have it make sense (cf. subheadings). Some citation styles write up names and dates in full, but this can get repetitious: "...Knuth [Knuth 83] has shown that ...". Don's paper on **goto**'s was published first in ACM *Computing Surveys* and later incorporated into a book. For this second printing he had to make numerous changes to the sentences containing citations, because the originals would look strange in the different context and format of the book. Oren Patashnik pointed out that the *Chigaco Manual of Style* recommends that you don't number references, lest you have to make changes all through the text every time you insert a new one. This is less of an issue when a system like TeX handles such things automatically, of course. The *CMS* is full of such efficiency tips.

Too many commas can be a bad thing (bad things?). For example, consider this sentence: "Our algorithm to recognize and label the graphs when given a directed graph, G, with distinguished vertex s, can be summarized as follows." Remove the commas around 'G' and put one after 'graphs'. As a rough guide, put a comma where a speaker would pause to draw breath.

The word "loop" was ambiguous when first used; Don replaced it by "self-loop".

A sentence in the paper began "If any H_j ($j > 0$) has ...". In fact it was known that H_0 satisfies the stated condition, so Don suggested that the authors simplify the statement by omitting the $j > 0$ condition. Moral: Give a simple rule rather than an optimal one.

Elsewhere we saw "...all the H_j's ...". This is of course the standard way to form the plural of a symbol, but you are going to get into trouble when you start also using the construct H_j' (that is, H_j primed). A simple way to avoid the problem is just to say: "...each H_j is ...". Alternatively, you might want to invent another name for the concept, particularly if you are going to be using it time and time again. It's just not elegant to have too many symbols crowded on the page. At one point the authors wrote "...of H_js descendants ...". This doesn't work at all; you do need an apostrophe for the genitive (possessive) case.

Three small points:

> Instead of "...the one vertex path ...", write "...the one-vertex path ...".

The preposition 'at' would be better than 'of' in "... vertex of distance $< d$ ...".

We certainly need a space here: "... using 4.3(below) we derive ...".

Some authors have a disconcerting habit of using a lemma or theorem that is not proved until later on in the book. This can leave the reader wondering whether someone hasn't pulled a fast one on him (essentially by using a result to prove that same result). So make it quite clear to the reader that your proof structures do respect the necessary partial ordering.

Using ties: TeX and other systems allow you to specify that certain blanks are not to be used as line breaks. For example, put ties after the word 'dominates' in the phrase 'v dominates x and x dominates w'. In "... if e is ..." it is best to put such a tie between e and 'is'. The idea is to keep line breaks from interrupting or distorting the message.

Beware of the unfortunate co-incidence! Sometimes we cannot use an idiom because some word is also being used in another sense. For example: "... n vertices have been deleted by this point." In one of the term papers, someone was using contour integration to study aerodynamics. There were airplanes and complex planes all over the place, much to Don's confusion. Another example that came up was the word 'left'. This can be either left as against right (in a tree structure, say), or a past tense of the verb 'to leave'. So 'the node x is left' might be ambiguous.

In a final remark for today, Don suggested that a sequence of examples that build upon one another is much more useful than a number of unrelated ones. The paper by Haddad and Schäffer has a particularly nice sequence of illustrations demonstrating this point.

After class, everyone got back two independently annotated copies of their term papers.

§38. Excerpts from class, December 2 [notes by TLL]

Don welcomed Mary-Claire van Leunen to her encore lecture by pointing out the intriguing books that Mary-Claire had placed on the desk; he said he hoped that she could now tell us all the things that the clock had prevented her from telling us last lecture. Mary-Claire countered by saying that Don had only invited her back "on the theory that no one could possibly be that nervous two weeks in a row."

Leaving the books alone for the moment, Mary-Claire told us the tale of *which* and *that*.

The story opens in the 17th century, when speakers of English have two relative pronouns: *which* and *that*. What are relative pronouns? Here are some sentences (examples taken from the *Concise Oxford Dictionary*) where *which* and *that* are used as relative pronouns, both singular and plural:

> Our Father which art in Heaven ...
> These are the ones which I want to learn.
>
> ... the one that I mean ...
> These are the ones that I want to learn.

Which and *that* are not always used as relative pronouns. Here are some sentences (again from the *Concise Oxford*) where they serve other functions (along with the technical term for the function they are serving):

Which? Say which.	(interrogative pronoun)
Which one? Say which one.	(interrogative adjective)
...during which time we can...	(relative adjective)
I like that.	(demonstrative pronoun)
I like that thing.	(demonstrative adjective)
...not all that wonderful...	(adverb)
...no doubt that he can...	(subordinating conjunction)
O that we could!	(particle)

We have no spoken evidence from the 17th century, but Language Theorists believe that writing and speech were very far apart. That is, they believe that no one's ideal was to write the way that he sounded. Theorists cite two pieces of evidence to support this claim: The first is that the Theorists themselves find it difficult to believe that, in the last three centuries, spoken English has evolved as fast as it must have if the written language and the spoken language originally matched. The second piece of evidence comes from examining extant 17th century guidelines on writing or speaking effectively.

By examining samples of writing from the 19th century (particularly the everyday writing that was used for communication rather than as examples of great literature), we can see that the written language has evolved into one much closer to the spoken language. Language Theoreticians of that time said that this evolution was good, but their admonitions came after the direction of evolution was already evident. (We should remember that our language belongs to millions of people. It cannot be controlled by the decrees of any one person or group.) We now move on to our own century.

In 1906 H. W. Fowler and F. G. Fowler published *The King's English* (Oxford University Press still has it in print). Here the brothers Fowler write down for the first time that conversational rhythms are to be reflected in written English.

In 1926 H. W. Fowler published *Modern English Usage* (also still available from Oxford University Press). In this book the surviving brother continues the explanation of the relationship between spoken and written English—but he does so much more clearly.

While we are following the hot trail of our current subject, we should not lose sight of the vast range of the contributions that Fowler made in this landmark book. Mary-Claire calls Fowler the "great theoretician of the semicolon." Fowler saw the semicolon, which has no spoken equivalent, as a structuring device that operates between the levels of the sentence and the paragraph. This is just one example of how Fowler tried to utilize the graphical nature of print to the advantage of written English.

Returning to the evolution of written English toward spoken English, let's examine how people use *which* and *that* when they talk.

Speakers do not use *which* as a relative pronoun because speakers do not normally express thoughts that are long enough to contain non-restrictive clauses: Our spoken sentences are shorter than our written ones. People do use *which* when they talk, but they use non-referential whiches to introduce new thoughts that are tacked on to old thoughts. Examples of this kind of usage seem strange when written down (because we don't use non-referential whiches in written English), but they sound perfectly normal when heard on the street. Here is one:

> I went sailing this weekend; which tells you why my nose is pink.

Fowler realized that written English would sound more like speech if the choice of relative pronoun was uniquely determined by whether or not the clause it introduced was restrictive or non-restrictive. He wrote several thousand words on this subject; here are a few of them:

> A supposed, and misleading, distinction is that 'that' is the colloquial and 'which' the literary relative. That is a false inference from an actual but misinterpreted fact. It is a fact that the proportion of 'that's to 'which's is far higher in speech than in writing; but the reason is not that the spoken 'that's are properly converted into written 'which's. It is that the kind of clause properly begun with 'which' is rare in speech with its short detached sentences, but very common in the more complex and continuous structure of writing, while the kind properly begun with 'that' is equally necessary in both. This false inference, however, tends to verify itself by persuading the writers who follow rules of thumb actually to change the original 'that' of their thoughts into a 'which' for presentation in print.
>
> The two kinds of relative clause, to one of which 'that' and to the other of which 'which' is appropriate, are the defining and the non-defining; and if writers would agree to regard 'that' as the defining relative pronoun, and 'which' as the non-defining, there would be much gain both in lucidity and in ease. Some there are who follow this principle now; but it would be idle to pretend that it is the practice either of most or of the best writers.

There is no doubt that Fowler has had a significant influence on the English language, but why is it that his effect on American English has been greater than on British English? To answer that question, we move our focus to New York in the year 1925: Harold Ross has just founded the *New Yorker* magazine.

Ross was a man who liked things to be clearly defined. He took *Modern English Usage* as gospel. For decades the *New Yorker* had reliably influential prose, and for decades H. W. Fowler's dictums were applied blindly to that prose. Mary-Claire was nearly nonplused as she mentioned reading a collection of letters from a *New Yorker* editor to various literary luminaries. ("I'm sorry, but we had to change all your *which*es to *that*s," sounds rather presumptuous when addressed to John Updike.) The *New Yorker* no longer treats Fowler as divinely inspired, and they haven't since the 1950s, but that leaves close to three decades of blind obedience to consider.

According to Mary-Claire, Harold Ross's attachment to obedience is not unusual—for Americans. She says that Americans look in a reference work, see what it says, and then decide to obey or to disobey; Britishers look in the same reference, see what it says, and then formulate new and different ways of treating the same questions. Why is it that Britishers feel that they are entitled to an opinion but Americans do not? Two partial answers might combine to give us a single satisfactory one.

Most British people have been English speakers for generations, but most Americans are descendants of recent immigrants. Immigrants are told, "These are the facts. If you want to speak English, follow the rules." Perhaps more important, educated British people are taught to write from day one. Many of the exercises that Mary-Claire gave us in her lecture on Calisthenics are actually used in British Grammar Schools. British University students discuss their weekly writing with their tutors—and they regularly write about 2000 words a week.

In 1957, we Americans acquired a new source of authority on writing English—this time *American* English. An old classmate of E. B. White's sent him a copy of the book that they had received from their English professor, Will Strunk. The decision-makers at Macmillan decided to publish a book that contained Strunk's monograph plus an extra chapter by White on "spiritual things." The combination of Strunk's clear and simple instructions and White's beautiful prose made *The Elements of Style,* by William Strunk, Jr., and E. B. White, the landmark of written style for our generation. Here is their entire essay on *which* and *that*:

> "That" is the defining or restrictive pronoun, "which" the non-defining or non-restrictive. See under Rule 3.
>
> > The lawn mower that is broken is in the garage. (Tells which one.)
> >
> > The lawn mower, which is broken, is in the garage. (Adds a fact about the only mower in question.)

Rule 3 says "Enclose parenthetic expressions between commas."

Mary-Claire has a copy of the first edition of *The Elements of Style,* in which White uses a 'which' for a 'that' (this has been changed in later editions). The line originally read:

> ...a coinage of his own which he felt was similar to...

The Elements of Style has many departures from guidelines presented by Fowler. It was written for the American audience, and it was written for an audience without a high level of grammatical sophistication. In contrast, as Mary-Claire said, "Fowler is rough going for those of us whose Latin is weak and whose Greek is non-existent." Future editions of Fowler may need prefaces explaining what adjectives, adverbs, and the like are. It is most common for people to learn those terms when they learn their first non-native language (though Latin is the only language to which the terms are perfectly suited).

Fortunately for native English speakers, there is a rule completely lacking in jargon that we can use to determine whether a 'which' should be a 'that':

> If you can substitute 'that' for 'which', do it.

Mary-Claire attributed this rule to Leslie Lamport. Leslie says that his version of the rule is actually:

> If it sounds all right to replace a 'which' by a 'that', then Strunk & White say replace it.

Which brings us to our next issue: Are *which*es that could be *that*s always wrong? Don said that now that he knows the rules, he finds every "wicked which" an irritating distraction from his reading enjoyment. He seemed to imply that *which*es in restrictive clauses are always wrong.

Mary-Claire said that the rules given hold for "everyday, expository prose—shirtsleeve prose, not literary prose." (She did not tell us how to decide when the everyday rules should be violated.) As for Don's discomfort on finding *which*es in well-beloved authors, she said "I believe we are encountering obedience here." After class, Leslie Lamport had this to say on the same subject:

> I unfortunately have somewhat the same reaction as Don to "incorrect" uses of *which*, for which I curse the evil influence of Strunk & White. When I observed that writers such as Dickens and Fowles are "incorrect," I quickly lost my desire to be "correct." However, I can't completely unlearn the reflex of being bothered by the "incorrect" usage.
>
> I still try to use *that*s when Strunk & White tells me to, because I know that many of my readers will have been similarly indoctrinated. But I will throw in an occasional wicked *which* to avoid a string of *that*s.

Mary-Claire's final word on the subject:

> "*Which* and *that* are not in themselves very important. But tone is important, and tone consists entirely of making these tiny, tiny choices. If you make enough of them wrong—choices like *which* versus *that*—then you won't get your maximum readership. The reader who has to read the stuff will go on reading it, but with less attention, less commitment than you want. And the reader who doesn't have to read will stop."

§39.   **Excerpts from class, December 2 (continued)**   [notes by TLL]

During the final moments of class, Mary-Claire finally got to those intriguing books. They were worth waiting for.

First, she reminded us of "Franklin's Exercise" from her previous lecture. She read us a passage from *The Autobiography of Benjamin Franklin* where Franklin mentions it:

> A question was once, somehow or other, started between Collins and me, of the propriety of educating the female sex in learning, and their abilities for study. He was of opinion that it was improper, and that they were naturally unequal to it. I took the contrary side, perhaps a little for dispute's sake. He was naturally more eloquent, had a ready plenty of words; and sometimes, as I thought, bore me down more by his fluency than by the strength of his reasons. As we parted without settling the point, and were not to see one another again for some time, I sat down to put my arguments in writing, which I copied fair and sent to him. He answered, and I replied. Three or four letters of a side had passed, when my father happened to find my papers and read them. Without entering into the discussion, he took occasion to talk to me about the manner of my writing; observed that, though I had the advantage of my antagonist in correct spelling and pointing (which I ow'd to the printing-house), I fell far short in elegance of expression, in method and in perspicuity, of which he convinced me by several instances. I saw the justice of his remarks, and thence grew more attentive to the manner in writing, and determined to endeavor an improvement.
>
> About this time I met with an odd volume of the *Spectator*. It was the third. I had never before seen any of them. I bought it, read it over and over, and was much delighted with it. I thought the writing excellent, and wished, if possible, to imitate it. With this view I took some of the papers, and, making short hints of the sentiment in each sentence, laid them by a few days, and then, without looking at the book, try'd to compleat the papers again, by expressing each hinted sentiment at length, and as fully as it had been expressed before, in any suitable words that should come to hand. Then I compared my *Spectator* with the original, discovered some of my faults, and corrected them. But I found I wanted a stock of words, or a readiness in recollecting and using them, which I thought I should have acquired before that time if I had gone on making verses; since the continual occasion for words of the same import, but of different length, to suit the measure, or of different sound for rhyme, would have laid me under a constant necessity of searching for variety, and also have tended to fix that variety in my mind, and make me master of it. Therefore I took some of the tales and turned them into verse; and, after a time, when I had pretty well forgotten the prose, turned them back again. I also sometimes jumbled my collections of hints into confusion, and after some weeks endeavored to reduce them into the best order, before I began to form the full sentences and compleat the paper. This was to teach me method in the arrangement of thoughts. By comparing my work afterwards with the original, I discovered many faults and amended them; but I sometimes had the pleasure of fancying that, in certain particulars of small

> import, I had been lucky enough to improve the method or the language, and this encouraged me to think I might possibly in time come to be a tolerable English writer, of which I was extreamly ambitious.

Next, she reminded us of the verse-writing exercises that she had so highly recommended during the previous lecture. She showed us the book that she and her student, Steven Astrachan, had worked through together: *A Prosody Handbook*, by Shapiro and Beum. She said what would have been even better was *Rhyme's Reason* by John Hollander. Mary-Claire said we should all buy Hollander's book, and then she tried to make sure we would by playing on the Computer Scientist's love of self-reference. This is Hollander's description of one particular poetic form:

> This form with two refrains in parallel?
> (Just watch the opening and the third line.)
> The repetitions build the villanelle.
>
> The subject established, it can swell
> across the poet-architect's design:
> This form with two refrains in parallel
>
> Must never make them jingle like a bell,
> Tuneful but empty, boring and benign;
> The repetitions build the villanelle
>
> By moving out beyond the tercet's cell
> (Though having two lone rhyme-sounds can confine
> This form). With two refrains in parallel
>
> A poem can find its way into a hell
> Of ingenuity to redesign
> the repetitions. Build the villanelle
>
> Till it has told the tale it has to tell;
> Then two refrains will finally intertwine.
> This form with two refrains in parallel
> The repetitions build: The Villanelle.

Mary-Claire told us that she once wrote out a recipe for making bagels in Alexandrine couplets. It was a good exercise, and it was hard. She says that it was so hard that she actually began to believe that the results would be intelligible (and interesting) to someone else. She sent the recipe off to a food magazine and received "a truly astounded letter of rejection." She cautioned us again that the verse exercises, useful as they are, "really are only exercises."

The final book that she showed us was *A Handlist of Rhetorical Terms*, by Richard A. Lanham. She said Lanham is the source for many of the great words she dazzles people with. Some of the terms in Lanham's book are more useful than others; there are some terms in the book that can only be represented in Greek syllabic verse.

Mary-Claire and Steven wrote out examples of each term in Lanham's book. Having performed the exercise, Mary-Claire confidently told us that it was not profitable. She said that her warning us not to try exercises that won't do us any good proves that she isn't totally crazy—and that the exercises that she did give us are worth doing.

§40. Excerpts from class, December 4 [notes by PMR]

> "All the officer patients in the ward were forced to censor letters written by all the enlisted-men patients, who were kept in residence in wards of their own. It was a monotonous job, and Yossarian was disappointed to learn that the lives of enlisted men were only slightly more interesting than the lives of officers. After the first day he had no curiosity at all. To break the monotony, he invented games. Death to all modifiers, he declared one day, and out of every letter that passed through his hands went every adverb and every adjective. The next day he made war on articles. He reached a much higher plane of creativity the following day when he blacked everything in the letters but *a*, *an* and *the*. That erected more dynamic intralinear tensions, he felt, and in just about every case left a message far more universal."
>
> — from *Catch-22* by Joseph Heller

Don rewarded today's early birds with the chance to participate in a referendum. We voted to decide the due-date for the term papers, Monday 14th or Wednesday 16th. UN observers were not surprised to find the latter date was favoured by the populace; the only surprise was that the vote was not quite unanimous. Very well then, said Don: All papers to be handed to him, his secretary or TA's, by 5pm (Pacific Standard Time) on Wednesday 16th December. (The *real* early birds were rewarded with some cookies that Sherry was handing round. And very good they were too).

It was of course too much to hope that we could get through the whole of a CS class without computers rearing their ugly heads; today they did. Don's topic was computer programs that are supposed to help us with our writing. Two such—`style` and `diction`—are available on Navajo (a CSD Unix machine). These are relatively old programs. State-of-the-art systems cost a lot of money, and so naturally Stanford doesn't have them. There is a program called `sexist`, for example, which attempts to alert us to controversial word usage. Don recalled the occasion when the (London) *Times* quoted him as saying that it wasn't appropriate to talk about 'mother and daughter' nodes in a tree structure, and he received a lot of irate mail as a result. People seem to be less uptight about such things these days, he said.

The `style` program takes a piece of text and scores it according to 'readability'. The analysis is very superficial—way below the level of human critiquing. However, said Don, these programs are kind of fun. And they do provide an excuse to read the document from another point of view. Even if the analysis is wrong it does prompt you to re-read your prose, and this has to be a good thing. Don recalled Richard Feynman's anecdote about his first day at Oak Ridge Laboratories: Having no idea what he was supposed to be doing, Feynman pointed to a random symbol in the blueprints and said, "What about this then?" A technician immediately agreed that Feynman had spotted a significant and potentially dangerous oversight in the design.

To illustrate the programs, Don had run them on a dozen or so sample texts. For instance, he used a passage from the rather ponderous introduction to a book by Alonzo Church; samples of PMR's and TLL's notes for CS 209; versions of his own exposition of binomial coefficients, vintage 1965 and 1985; *Wuthering Heights*; *Grimm's Fairy Tales*; and part of a book about the Bible that Don is writing on weekends. The `style` routine produces four different readability grades for any piece of text. Each is literally a "grade" in that it indicates what level of education the piece suggests. The basis of the grading is very straightforward; it's a linear formula whose variables are the average number of syllables (or letters) per word and the average number of words per sentence (or sometimes the reciprocal of this value). For example, there are constants α, β, γ such that

$$\text{grade} = \alpha \text{ (words/sentence)} + \beta \text{ (syllables/word)} + \gamma.$$

How were α, β, and γ determined? The authors of each readability index simply look at a large number of pieces of writing and assign them a grade-level 'by eye'—that is, they estimate the age of the intended reader. Each piece of text is then characterised by three real numbers: the average number of words per sentence, the average number of syllables per word, and the subjective grade level. So each piece determines a single point in 3-space (plotted against three orthogonal axes); the set of pieces determines a scatter of points in 3-space. Standard linear regression techniques are used to find the plane that is the "best fit" for these points. The three parameters above define this plane.

Someone asked whether we should be shooting for some specific grade level, and if so, what level? Don replied that his usual aim is to minimise the level, although overdoing this will defeat the purpose.

In addition to the raw scores, a variety of other parameters come out of a `style` analysis: average length of sentences, percentage of sentences that are much shorter or longer than the average, percentage of sentences that begin with various parts of speech, etc. The program also attempts to classify sentences into types and tabulate their frequencies, as well as telling us the percentages of nouns, adjectives, verbs (active or passive), etc. A sentence is considered "passive" if a passive verb appears in it anywhere, even in a subclause. Curiously, `style` classifies any sentence that begins 'It ...' or 'There ...' as an "expletive." This seems a little strange to those of us who are old enough to remember Watergate. We always thought that it was quite a different class of words that the transcribers of Tricky Dicky's tapes felt the need to delete.

Don's theological piece stood out as being pitched at a significantly lower grade level than the other specimens. He was initially surprised by this, and double checked the data to make sure there was no mistake. But on reflection he concluded that we usually write more obscurely when writing about our own field. The two versions of his binomial chapter had very similar scores, despite their having been written twenty years apart. Church's piece scored high. Don said that the statistics were misleading here; although Church's sentences are quite long, they are not ugly but musical. Still they were not a special joy for the reader.

The `style` output also noted a lot of passive voice in Church (perhaps not surprising in a technical work) and a paucity of adjectives in *Grimm's Fairy Tales*. Don noted that Mark Twain didn't think much of adjectives either.

A companion program called `diction` operates on different lines. It has an internal dictionary of 450 words and phrases that it deems 'questionable' and flags them, inviting the writer to find an alternative way to express himself. For example, `diction` doesn't like the word 'gratuitous', and flags its use as an error. Neither does it like the phrases 'number of' or 'due to'. Don noted that copy editors generally prefer 'because of' to 'due to' in ordinary writing, and perhaps `diction` is overlooking the mathematical usage: "This theorem, due to Cauchy, is used ...". In Don's book T_EX: *The Program*, the copy-editor changed all Don's 'due to's to 'owing to'; Don changed them all back again. But he searched unsuccessfully for a reference to the mathematical usage in his dictionaries, so he wondered aloud if he was completely out of line with the rest of the world. The class unanimously reassured him that 'due to' was quite the elegant way to give credit for a scientific innovation. Lexicographers are out of touch here.

The word 'very' is also on `diction`'s list of suspects. Don recalled that someone had once advised him thus: "Try changing all your 'very's to 'damn's and see what results. Don't use 'very' unless you would happily use 'damn' in its place." Damn good advice!

The `diction` filter also objected to 'literally' and 'in fact', but partially redeemed itself by catching a wicked 'which'. A sister program, `explain`, expands on `diction`'s objections and recommends improvements. For example, `explain` suggests that we write 'if' instead of 'assuming that', and 'really' instead of 'actually'. In practice, users reportedly accept about 50% of `diction`'s suggestions. And that's as it should be—we've got to keep these machines in their place.

§41. Excerpts from class, December 7 [notes by PMR]

Today we heard from our penultimate guest speaker, Rosalie Stemer. Rosalie is a wire features editor at the *San Francisco Chronicle*, teaches copy editing at Berkeley, and has worked as a copy editor for the *San Francisco Chronicle*, the *Kansas City Star*, and *Chicago Daily News*. So she wields an ultimate pen.

It's a sad truth, Rosalie said, that people who should be able to write well often can't. She illustrated with a newspaper headline:

> DISABLED FLY TO SEE CARTER

and a story that began: "Doing what he loved best, golf pro John Smith died while ...". She told us about the occasion when a newspaper was having trouble fitting the word 'psychiatrist' into a headline, and resolved the problem simply by writing 'dentist' instead.

Rosalie went through a story filed by an experienced journalist, pointing out its good and bad features and the changes she had made.

> "Nine out of ten books bought in this century by the U.S. Library of Congress, one of the great research libraries of the world, will self-destruct in 30 to 50 years."

She faulted this sentence on a number of counts. The 'great research libraries' phrase puts the wrong focus on the sentence—we are not really concerned about the status of the Library of Congress in this article. 'Nine of ten ...' would be better, she said; the word 'out' is superfluous. And does the Library of Congress buy the books that it houses? No. Publishers *give* books to the Library of Congress, as required by law.

The second sentence noted that the problem is plaguing fine book collectors, among others. What does 'fine' modify here, Rosalie asked: the books or their collectors?

> "Yet many books several hundred years old are in excellent condition, Dr. Norman Shaffer, the congressional library's director of preservation, said yesterday."

Rosalie thought the subject and verb too far apart. Moreover, she said, it raises the question: "*Why* are they in excellent condition?"

Other points: "... the cheaper process of making their products ..."—cheaper than what? "Another solution lies in persuading ..."—a solution to what? And who should be doing the persuading? Rosalie saw a systematic error here. A hallmark of good writing is that it answers more questions than it raises, she said.

Someone asked whether reporters perhaps write a little sloppily in the knowledge that the copy-editor will go over their copy and clean it up? Rosalie said that they certainly *shouldn't* do this. Someone else pointed out that it's probably a bad idea to start talking about solutions (to problems) in the same breath as *chemical* solutions. This is the "unfortunate coincidence" problem that Don talked about recently.

Rosalie shuddered over an extremely awkward sentence about acids and alkalis—fortunately someone in the class was able to decipher and explain it. "One hopeful sign ..." was another problem—the sign is not full of hope. Over and over again we saw sentences in which the subject and verb are separated by many words—these are hard to read. Rosalie pointed out a number of places in which whole phrases could be dropped without any loss of meaning: "Dr. Shaffer said one of the most encouraging signs is the fact he has heard one of the largest paper manufacturers ..." can be reduced to "Dr. Shaffer said one of the largest paper manufacturers ...". Whenever you see "he has heard" you can often improve or delete it, Rosalie said. A similar case: "The reason for removing the spaces from the list is that ..." can be (better) written "We remove the spaces because ...".

Rosalie conceded that good writing is very difficult. We must strive to be clear, coherent, accurate, and concise. This last is especially important, she said, and quoted Pascal: "I have made this letter longer than usual because I lack the time to make it shorter."

Rosalie was pleased to note that the first drafts of our term papers were quite a bit better than something else she had read recently—a report by a local software company. After just a few weeks we are pushing out the envelope of Silicon Valley literacy! But many of our sentences could be improved, she said, by cutting them shorter. Out with the semicolons, in with the periods. Don't write one long sentence if you can say the same thing in two short ones. A semicolon should be used only where the separated clauses have a very close relationship, and even there a period is often better. She quoted William Zinsser in his book *On Writing Well* as saying "The semicolon all too easily conveys 'a certain

19th-century mustiness' and slows the pace of the writing." Another common error was the frequent repetition of a word like 'this', 'they', 'just', or 'then'. Reading the piece aloud will often help you spot such over-uses.

Rosalie wasn't too keen on some of the conversational idioms that crept into our writing: sentences that begin 'Anyway, ...'; an algorithm described as 'pretty straightforward' (perhaps a bad idea in a paper on pretty-printers?). Neither did she like the phrase 'again iterate through'—this sounds awkward; surely the same point could be put more smoothly? A lot of sentences suffered from not having their subject near the beginning: "If ..., the graphic interface ...". Someone suggested that these kinds of problems—as well as over-use of the passive voice—are easily avoided if we stick to a subject-verb-object style.

Someone asked whether 'in the context of' wasn't a "noise-phrase"—one that could be deleted without any loss of meaning? Rosalie said that this was often so, but that sometimes it can mean something. In the example we were looking at, the phrase had been used early on, so it seemed reasonable to repeat it somewhere else to make it clear to the reader that we were once again talking about the same thing as before.

A student asked whether perhaps there aren't different styles of writing appropriate to scientific journals and to newspapers? Newspapers would probably put a greater premium on simple, direct sentences, for example. Rosalie said that there might be something in this, but that clear writing was always good.

The sentence "Each of the Δz's are then multiplied by this factor" can be improved on two counts: Change 'are' to 'is' and eliminate the passive voice. How about: "Multiply each Δz by this factor"?

If you are using commas to insert a parenthetical note, you must put a comma on *both* sides: "...this node, b, is thus ...".

Rosalie didn't like a sentence that began 'It is unlikely ...'. The pronoun reference problem appears: *What* is unlikely? There is no uniform rewrite-rule for this, but we can usually find an alternative construction that conveys the same meaning. Like Jeff Ullman, Rosalie wasn't enthusiastic about 'This is done by ...'; better, she said, to say "This procedure (process, step, etc.) is done by ...".

A very common error is the misplaced 'only'. To illustrate, Rosalie took the sentence "I hit him in the eye yesterday" and inserted 'only' in each of the eight possible positions. Sure enough, each resulting sentence carries a somewhat different force:

> Only I hit him in the eye yesterday.
> I only hit him in the eye yesterday.
> I hit only him in the eye yesterday.
> I hit him only in the eye yesterday.
> I hit him in only the eye yesterday.
> I hit him in the only eye yesterday.
> I hit him in the eye only yesterday.
> I hit him in the eye yesterday only.

If you say "Here we only calculate the position of two vertices" you probably mean "Here we calculate the position of only two vertices".

We saw a sentence that contained four or five occurrences of the word 'then'—surely a trifle excessive? Someone remarked that the sentence was probably an anglicised version of a line of computer code, which abounds in '**if** ... **then** ...'s, sometimes deeply nested.

Another line that caught Rosalie's eye was: "...saving the computation for the place where it is really needed." The word 'really' was used again the same paragraph. She thought this sounded altogether too vague for a piece of technical writing: Is the computation needed or not? What is this "really needed"? There was a definite difference of opinion over this question: Some people in the class couldn't see any objection to this usage. Someone argued that the "really" amounted to stylistic advice (as in "when painting a house, be especially careful on the window-frames, where precision is really important"), but it is by no means superfluous—the word makes a substantive contribution to the meaning of the sentence. Rosalie's objection stemmed mainly from the fact that the word "really" is much over-used in colloquial speech. In the end we agreed that it would probably be better to say something like: "...saving the computation for those vertices where the additional work contributes more to the visual quality".

Can an object witness a property? To Rosalie's ear this was a strange construction. But the class assured her that this is common usage in computer science. Technical terms take on an anarchic life of their own!

In the last minute, Rosalie showed us a list of pairs of words that are frequently—and sometimes amusingly—conflated. For example, 'prostrate' and 'prostate'. One common confusion is 'alternately' vs. 'alternatively'. These are not synonyms. (Alternately, Tracy and I take notes in class. You could read them, or alternatively you could take your own notes.)

§42. Excerpts from class, December 9 [notes by TLL]

Don started class by introducing Paul Halmos. Paul is a distinguished author, a professor of mathematics at the University of Santa Clara, and a spicy and entertaining lecturer. As Don said, "He brings our program of guest speakers to a triumphant conclusion."

Paul started his lecture by wondering why we had called him here. "I don't have anything new to say," he said. "What I had to say has already been majorized by Don and Mary-Claire." He said that even the act of talking about mathematical writing was difficult, by comparison with the act of talking about mathematics itself. We don't have to remember much about math, because we know its structure; we can develop and discover the material as we talk about it. The structure of mathematical writing is much more elusive, so how do we know what to say about it? Sure, Paul brought several pages of prepared notes to class, but he claims that even those won't help him much.

Not that the subject of mathematical writing isn't important. Some mathematicians have disdain for anything other than great theorems. "Anything else is beneath them." But they are wrong. Mathematicians who merely *think* great theorems have no more done their job than painters who merely *think* great paintings.

Paul has read our handouts, and he wants to make a few comments. He wants to have a dialog with us; he admonished us to break in whenever we feel the urge.

He is going to drift in and out of many different topics but only after he has given us an anchor and a rough outline. The anchor? Two basic rules:

> Do organize material.
> Do not distract the reader.

The outline? Four aspects of good mathematical communication:

> Semantics (words, and the job they do);
> Syntax (also known as grammar);
> Symbols (very meaningful to mathematical writers);
> Style (synthesis of the above).

Turning first to Semantics, Paul spoke to us about the natural process of change inherent in language and how it affects our word usage. Some changes are good—some changes are bad. According to Paul, one of the most often discussed symptoms of that change is the word 'hopefully'.

The most recent literary tradition, handed down to us by our grandparents, tells us that 'hopefully' means the exact opposite of 'hopelessly':

> "I don't have a chance in the world to be promoted," he growled hopelessly.
> "My chances look good," his colleague grinned hopefully.

But another, impersonal use of 'hopefully' has become popular—an evasive form in which one can say "Hopefully he won't be re-elected" instead of "I hope he won't." This conflicts with the normal usage of other words that can end both -fully and -lessly. Although we may think that interest rates will rise, we don't say "Thoughtfully interest rates will rise."

Although we may fear that muggers are in the street, we don't say "Fearfully those muggers are still out there." Consistent English usage would prohibit

> Hopefully I'll visit you again next year,

as much as it prohibits

> Hopelessly I'll not be able to come.

Paul doesn't like the new usage, which he calls "illogical and ugly." The mere fact of change is bothersome. But he realizes that his is only one vote, and he seems to be outnumbered. On balance, it is perhaps a good change, one that might even make communication easier. "The English language won't collapse if the other side wins." In fact, Paul says,

> Arguably the change is a needed one.

But he is surprised to hear himself saying that.

Paul sees other changes as needless and careless. It grates on his ears when he hears,

> The earthquake decimated more than half the houses.

Of course some would say, Why do we need to reserve a special word for the random destruction of one out of ten? Paul thinks muddying the meaning of the word is bad, but he admits that it is harmless.

Other unneeded, and harmful, obfuscations should be discouraged. 'Imply' does not mean 'infer', and 'disinterested' does not mean 'uninterested'. To confuse these words is to lose valuable distinctions. Tragically, the differences between these words are becoming so confused that if we are writing for a large audience, and if we need to make use of the distinctions, we probably shouldn't.

Evidence of bad changes can even be found in our handouts. In §4, one of the TAs (not the one with the charming British Accent) used 'reference' as a verb. Paul's response: "There is no such verb, and if there were, it sure as hell wouldn't be transitive." How would it sound to say "I quotationed the author"?

Barry Hayes pointed out that in Computer Science, 'reference' is a technical term used as a verb. Technical terms like 'majorize' sometimes creep into our vocabularies. Don supported him by saying that computer programmers "reference and de-reference things all the time." Paul's response: "My condolences. You know, the French say English is ruining their language. How the French feel about English is how I feel about that."

We moved on to Syntax. "Obviously," said Paul, "people approve of it; nobody uses ungrammatical English on purpose." Syntax changes more slowly than semantics. However, he once heard the following lovely sentence:

> If I'da knowed I coulda rode, I woulda went.

This has rhythm, it's communicative, it's personal; but of course it's not grammatical English. Therefore it distracts the reader from what is actually being said. Here's another non-made-up example:

> Us'll go along with she if her'll go along with we.

If we are trying to communicate with people who use such grammar, we should use their language so as not to distract *them* from what we're saying. But as technical writers we are presumably not addressing that audience, certainly not in print.

Paul would like to advance the thesis that grammar is logic. This notion is abhorrent to linguists, who see grammar as illogical, inconsistent, and contradictory; and they are right. Nevertheless, grammar is the organizational principle that lies behind linguistic communication. A typical English sentence like 'He saw her' contains case, tense, and gender; such things give a tremendous amount of information in condensed form, and they can be seen as logic. To identify grammar with logic is less of an error than to reject logic altogether.

Speaking of case, Paul says, "Cases are good things, even though in English by now they are vestigial." They do exist, and they must be treated with respect. We say

> I don't know him,

but we wouldn't be caught dead saying

> I don't know he.

Similarly, we say

> He is the President of France,

but never

> Him is the President of France.

Therefore we would not logically ask,

> Whom is the President of France?

Simple, right? Well, there are more confusing cases too:

> I don't know who is the President of France.

Or should it be 'I don't know whom is the President of France'? A grammatical push-pull is involved here. (The nominative wins, and 'Who' is correct.)

Paul would like to stamp out abuses such as 'I hate whomever said that'. An attention to logical rules of grammar helps us to clarify our own thinking in general.

Taking issue with part of our first handout, Paul says the rule "A preposition is a bad word to end a sentence with" is "reactionary grammarian balderdash". Consider:

> Palo Alto is a good place to live in.
> Don Knuth is fun to have a drink with.
> There aren't many people I would say that to.

All of these are examples of prepositions in "post position" that could only be ruined by being made grammatically pure. (We have all heard Winston Churchill's famous statement about "the sort of nonsense up with which I will not put.") Why should we do gymnastics for sentences with only one preposition at the end? Paul gave us a famous sentence ending with five prepositions:

> What did you want to bring that book I didn't want to be read to out of up for?

On the discussed and re-discussed subject of 'which' and 'that', he says that Mary-Claire stole his thunder. It is worthwhile to get it right, but it is not terribly important.

We began discussing Symbols by discussing punctuation. Paul urges everyone (contrary to rule #25 in §1) to place quotation marks logically, every time. He gave us what he sees as a ridiculous example from Kate Turabian, whom he calls "The Antichrist," in *The Chicago Manual of Style*:

> See the section on "Quotations," which may be found elsewhere in this volume.

Paul was incensed. "Horrors", he said. "You see the illogic, don't you? There's no reason for it. It's not a grammatical convention—it's a totally arbitrary typographer's convention. The battle against this sort of stupidity can be won." He has succeeded in getting his own books punctuated logically. Bob Floyd gave support by mentioning how deadly such conventions are in a book about computer programming.

But then Don remarked that one of Paul's two main points was not to distract the reader. Paul said, "And your implied, snide, argument?" "Well," said Don, "I guess I'm implying that you think you're distracting only the copy editors and not the readers." "Yes, I believe that's right, with respect to commas and quotation marks."

Mary-Claire asked, "Just how far are you willing to go in the direction of logic? Are you willing to place periods outside the quotation marks in actual dialog that already has its own punctuation?". Her example:

> He said, "No.".

Paul said that if you push him in a corner he might go so fas as to say "Yes.". And Mary-Claire responded, "That's what I thought. Luckily there's not much dialog in the sort of stuff you write.". (Paul conceded that he doesn't really have an ear for dialog and doesn't have immediate plans to break into the world of fiction. He would love to write a novel, some piece of literature that isn't expository, but he's not being held back by an inability to punctuate.)

The second Symbols-related point that Paul wanted to bring to our attention was the subject of written versus symbolic numerals. He gave us an examples where 'one' could either be a pronoun or a numeral, depending on the context:

> What are we to do when x is one?

The sentence preceding that one may have been

> The solutions of the equation are the singularities of the function we are studying.

Or it may have been

> Everything is clear when x is 2 or greater.

Another example (this time from Birkhoff & MacLane's classic text):

> The first few positive primes are
>
> $2, 3, 5, 7, 11, \ldots$.
>
> Any positive integer which is not one

> or a prime can be factored ...

He urges us to remove such ambiguities by using '1' when we want to speak of the numeral.

> The number of solutions is either two or three.
> The only solutions are 2 and 3.

Paul now moved on to the final area of discussion: Style.

Rule #6 in §1 suggests that we use 'we' to avoid passive voice. This use of 'we' is equivalent to "the reader and I". Paul says that even better is to avoid both passive voice and the use of 'we' through judicial use of imperative and indicative moods along with an outlying kind of non-sentential phrase. For example,

> We can now prove the following result:

becomes

> A consequence of all this is the following result.

Or,

> Consequence: A implies B.

The latter technique can occasionally be used in a sequential manner,

> Consequence 1: X. Consequence 2: Y.

ending with a final blaze of glory,

> Conclusion: Z.

Alternatively, here's an example of imperative mood:

> All we need to do to get the answer is to replace x by 7 throughout.

Just say

> Replace x by 7 throughout.

Paul finds this less distracting. Using 'we' is not a crime, but it adds an irrelevant dimension that can often be replaced by something clearer and smoother.

He gave a lengthier example of a typical passage that shows how both 'we' and passive voice can be avoided without sounding artificial:

> If U is ⟨something⟩, then the spectral theorem justifies the assumption ⟨something⟩. If f is ⟨something⟩, then ⟨something⟩ equals ⟨something⟩. Since, however, y is ⟨something⟩, it follows that* ⟨something⟩. Since, moreover, the assumption ⟨something⟩ implies that ⟨something⟩, the Lebesgue theorem is applicable. This completes the proof of convergence.

(The example would be more effective, of course, if the ⟨something⟩s were replaced by meaty concepts, but that would distract us from the point at issue.)

---

* Passive, God forgive me, or at least not active; but this phrase is standard and inoffensive to my ear.

An audience member asked if using 'we' introduced a light tone that imperative doesn't have. Paul agreed that it does, and stated that he isn't sure he wants that tone in his writing.

Another questioner asked about first person singular? Paul likes it, but he admits that it can be disturbing: "Who does that jerk think he is?" He reluctantly agrees that the first person singular should be avoided in formal technical writing.

Leslie Lamport asked at this point if the use of 'we' could not be avoided by avoiding prose proofs in favor of tabular proofs. Paul didn't like that idea at all: He finds symbols insidious and much prefers prose proofs. But then he had second thoughts, saying that he and Lamport might not disagree too much on the need to rethink the techniques of proof presentation. Outline form (not too heavily symbolic) might be advantageous.

Paul said that he casts all possible votes in favor of Rule #9 in §1: Do not echo unusual words. We had been told that this is a good idea because it avoids monotony. Paul says that it is a good idea because two uses of the same word in unrelated passages will be associated in the reader's mind and cause unwarranted connections. (Bob Floyd says that it will also cause technical typists to omit all words between the two occurrences.)

Paul's next bugaboo ("Do I dare do this thing?") was the phrase 'he or she' when he feels the traditional neuter pronoun 'he' would be sufficient. As soon as he brought this up, Mary-Claire disagreed, but Paul held the floor and quoted from authority by reading Mary-Claire's words from page 4 of her own book:

> This 'his' is generic, not gendered. 'His or her' becomes clumsy with repetition and suggests that 'his' alone elsewhere is masculine, which it isn't. 'Her' alone draws attention to itself and distracts from the topic at hand.

Mary-Claire responded, "Deeply moving quotation, but it is not true that the traditional solution to this problem in English is 'he'. The traditional solution is 'they'."

Many people in the audience stated pieces of opinions, but time was nearly up. "To each their own." Paul moved to the next topic: Proof by contradiction. He emphasized that proofs by contradiction should not be used if a direct proof is available. For example, he noted that proofs of linear independence often say, "Suppose the variables are linearly dependent. Then there are coefficients, not all zero, such that ...contradicting the assumption that the coefficients are nonzero." This circuitous route can usually be replaced by a direct argument: "If the linear relation ... holds, the coefficients are all zero. Hence the variables are linearly independent."

Don pointed out that proof by contradiction is often the easiest way to prove something when you're first solving a problem for yourself, but such stream-of-consciousness proofs don't usually lead to the best exposition.

Paul wound up his speech by repeating his opening rules: "Do organize," and "Do not distract."

The trouble is that it is hard to say what organization is. But we recognize it when we see it. "Give me a book, or a paper or a manuscript, and I'll tell you if it is organized," said

The material is in linear order, but organization means much more than that. "The plot of an exposition is rarely a straight line." Branches and alternative threads must be woven together. Paul says he spends most of his writing time working on organization of the material. He suggests that we look at Roget's *Thesaurus*, an encyclopedia, a do-it-yourself article, and a good textbook, for increasingly complex examples of non-linearly-organized presentations.

"Do organize," and "Do not distract." Except that all rules are made to be broken. When you want to jar your readers, Paul suggests that you distract them by changing your notation, screaming ungrammatical sentences, or being awkwardly repetitious.

His final words to the class were, "Anything that helps communication is good. Anything that hurts is bad. And that's all I have to say."

§43. Excerpts from class, December 11 [notes by TLL]

The final lecture of CS 209 was partially devoted to course evaluation. (We were, no doubt, harsh but fair.) Don told us that we would spend the last 40 minutes of class looking at the notes of people who have been going over our handouts but haven't had a chance to speak. (More course evaluations, perhaps?) Don said that he wanted to "end on a note of honesty and truth."

The first comments that he addressed were from Nelson Blachman (father of course member Nancy Blachman). Nelson is very interested in writing (he writes papers frequently), and he took the time to suggest improvements to the first few handouts.

Don liked some of these suggestions, but he found others incompatible with his personal style. He said, "The main thing that I get from this is that the style has to be your own. You will write things that someone else will never write." Don says he has learned this lesson well by writing an annual Christmas letter with his wife, Jill. "We get along 364 days of the year," he said, "but there is no way that we can write a sentence acceptable to both of us." (They have solved the problem by writing alternate paragraphs.)

Among Nelson's suggestions were:

> Changing 'the above proof' to 'the proof above'. Don agrees with this change mostly because editors are forever calling him on it, but the original usage doesn't sound terribly odd to him. Nelson says that 'above' and 'below' are two adjectives that never precede the things they modify. Don thinks 'above' has become an adjective, but 'below' hasn't (yet).

> Changing ', i.e.' to '; i.e.'. Don says that that is a matter of taste and pacing.

> Changing the spelling of 'hiccups' to 'hiccoughs'. Don's dictionaries preferred the shorter spelling.

> Changing 'depending on the usage, the terms this, that, or the other might be used' to 'depending on the usage, the term this, that, or the other might be used'. Don didn't see this as an improvement.

Changing 'programming language notation' to 'programming-language notation'. Don said that the suggestion might be appropriate for readers in other disciplines, but in our field the hyphenation would become annoying. Analogous cases are 'random number generator' and 'floating point arithmetic', each of which is potentially ambiguous, but so familiar in computer science that a hyphen looks wrong.

Then Don briefly showed us an example of a problem that often occurs when mathematicians are allowed to typeset their own text. A novice typesetter tends to make fractions like $\frac{n(n+1)(2n+1)}{3}$ instead of using the more readable slashed form $n(n+1)(2n+1)/3$.

Next, we returned to Mary-Claire's essay on 'hopefully' (see §26 above). Don says that he passed it out to us more for the style of the essay than the content, but it does make good technical points as well. To his surprise, Mary-Claire said that after re-reading it she actually wanted to improve the style. (This proves once again that nothing is perfect.) Here is what Mary-Claire wrote to him:

1) The dates should be expressed in the same terms. Given that I'm going to need to say '1637', I have to say 'late in the 1500s', not 'late in the 16th century'.

2) The sentence

> Impersonal substantives, on the other hand, serve less often than personal ones at the head of the kind of active verbs we modify with adverbs of manner

is so horrid I'd prefer to think I was drunk when I wrote it. To fix it I have to rewrite the whole paragraph, sliding 'adverb of manner' up earlier:

> As with most adjectives, both of these 'hopeful's regularly produced '-ly' adverbs of manner. The kind of hopefulness that means expectant and eager produced adverbs more readily than the kind that means promising and bright. There's nothing mysterious about that difference in frequency. The pattern
>
> ⟨personal noun⟩ ⟨active verb⟩ ⟨adverb of manner⟩
>
> is very common. People can carry themselves hopefully or eye a desirable object hopefully or prepare themselves hopefully for a possible future. The pattern
>
> ⟨impersonal noun⟩⟨active verb⟩⟨adverb of manner⟩
>
> is less common. Impersonal nouns serve less often than personal ones as subjects of the kind of active verbs that we modify with adverbs of manner. Nonetheless, a wager can be shaping up hopefully, a day can begin hopefully, ...etc.

Bob Floyd sent a few comments to Don, beginning with his opinion of the usage of hopefully. First, he reports that only 44% of the American Heritage Usage Panel found the use of hopefully as a sentential adverb acceptable. Bob also provided several authoritative quotations to support his objection to its use. (Don said that this is the main concern: Using 'hopefully' raises hackles in many people, distracting them from what you're trying to say; that's why he doesn't use it. But he thinks some of the documents that Bob uses to support his position were probably written by the the people that Mary-Claire was calling ignorant in her essay.)

Tom Henzinger, who is Austrian, observed that the German language has a common word 'höffentlich' that corresponds precisely to the new English usage of 'hopefully'. This reminded Don that he often needs words that the English language just doesn't have. For example, we have hundreds of ways to say that Jane beat Jim, but we have few ways to say that Jim lost to Jane. (And we have to use two words in the latter case where only one is needed in the former.) Don said:

> Our language often lacks verbs that correspond to "reflexive" relations. We have an abundance of words like 'dominate' but none like 'dominate or equal to'. So we must use long-winded phrases like 'less than or equal to'; sometimes, but not often enough, we can say 'at most'.

Returning to Bob Floyd's comments, Bob sent Don several citations to support his claim that exclamation points should be used only with actual exclamations or interjections. Some examples: Ouch! Stop! Thief! Well, I'll be! To Don's surprise, none of the authorities even mention that exclamation points can indicate surprise! Paul Halmos, speaking from the peanut gallery today, told about a trick he has to get around this: You can put the exclamation point in parentheses(!).* Then everybody is happy, because you've made an exclamation of surprise.

Bob said, "Advice to always avoid splitting infinitives is unwise." Don agreed that split infinitives can provide good emphasis and that rewrites can sound forced or awkward.

About not ending sentences with prepositions, Bob said, "You have no case, give up." Don agreed, saying that he had not understood the issue. "Coming from Milwaukee, where half the people speak English with a heavy dose of German, has made me oversensitive to sentences that end funny." However, there is a problem with sentences ending with prepositions, namely when they already have a structure that accommodates the preposition in the middle:

> Avoid such prepositions, which such sentences end with. The people who don't like the rule against prepositions in post position would never think of writing such sentences, so they probably have forgotten why the overly restrictive rule was first formulated.

Bob next objected to Don's suggestion not to omit 'that's. Don admitted that there are cases when leaving out a that produces a better sentence. For example, 'He said he was

---

  * Don was able to use that trick the next day in Chapter 8 of his book. (Who said this course wasn't practical?) But he found that it was like an unusual word: You can't easily repeat it twice in the same chapter.

going' is a better sentence than 'He said that he was going.' But, in this example 'that' is not needed as a grammatical help because the pronoun (in nominative case) keeps the syntax clear. In technical writing we often have more complicated sentences, which can benefit from the extra information that 'that' provides.

Someone in the class mentioned a related issue: Should the word 'then' be used in sentences like "If I get there early enough, ⟨then⟩ I will save you a seat." (Rosalie had suggested that it should not.) Don says that there is a difference between technical writing and newspaper writing, and he believes that well placed 'then's can make a paper more easily understood. In that particular sentence he would definitely leave out 'then'; but in mathematical contexts (where the phrase after the comma is likely a mathematical statement) he would definitely leave it in. Don says that our brains only have time to do simple parsing when we are reading for speed and comprehension. As Paul Halmos said, "Anything that helps communication is good."

The final subject that Don introduced was a behind-the-scenes discussion between Mary-Claire, Don, and one of the class members: Dan Schroeder. Dan received the comments on his term paper and objected to the claim that he had "wicked-whiches"; he gave involved logical reasoning in support of why his whiches really should be whiches. Don said, "If you have to think that long about the sentence, it is probably wrong." Mary-Claire said that writers have to contend with overly-sensitive readers like Don, who wince at all whiches that aren't preceded by commas or prepositions.

In one place Dan did not place a comma before a which because he was concerned about coincident commas. This is what Mary-Claire has to say about coincident commas:

> Coincident commas are not a sign of bad construction, any more than the coincidence of a final comma and a period, or a final comma and a semicolon, or any other two marks of punctuation. Where two commas coincide, we write only one. Where a comma and a period coincide, we write the period. Etc. Truly, coincident punctuation is not a problem.

(Did you catch the coincident periods there?)

After this comment we were thrown from the room in order to make way for another class. As always in this course, there was more to say than there was time to say it in.

Postscript: The instructor received an anonymous contribution after class, in response to his request for a poetically stated computer program:
```
      This algorithm to count bits
    Rotates VALUE one left and sums its
        two's-comp negation
        in a zeroed location
    Repeats WORDLENGTH times, then exits.
```

(Not only does this rhyme and scan, it also works. In fact, it may be the fastest way to do sideways addition on the GE635 and similar machines.)